A NATUR

DANGEROUS
CREATURES
OF
AUSTRALIA

Peter Rowland & Scott Eipper

JOHN BEAUFOY PUBLISHING

First published in the United Kingdom in 2018 by John Beaufoy Publishing Ltd
11 Blenheim Court, 316 Woodstock Road, Oxford OX2 7NS, England
www.johnbeaufoy.com

Photo Credits
Front cover: *main image* Eastern Mouse Spider © www.kapeimages.com/Peter Rowland,
bottom left: Coastal Taipan © Scott Eipper/Nature 4 You, *bottom centre:* Honey Bee © www.kapeimages.com/Peter
Rowland, *bottom right* Bullrout © Jason Sulda
Back cover Southern Blue-ringed Octopus © PT Hirschfield/Pink Tank Scuba (www.pinktankscuba.com).
Title page Inland Taipan © Scott Eipper/Nature 4 You.
Contents page Great White Shark © Elias Levy.
Main descriptions Photographs are denoted by a page number followed by t (top), b (bottom), l (left), c (centre) or r (right).

Thomas Alexander 86t; Luke Allen 143t; Amanda 44 88t; Deb Aston/astonunderwaterimages.com 116; Australian
Institute of Marine Science (AIMS) 24t; Daiju Azuma 26b; John Barclay 89t; Silke Baron 106t; Doug Beckers 43t;
Philippe Bourjon 103t; Thomas Bresson 66t; Brisbane City Council 157b; Andrew Brown 72t; Steve Bullock 81t; David
Burdick 117b; Brian Bush 145cl&cr; Center for Disease Control and Prevention 65t, 80b; Hal Cogger 152bl&br, 153b;
Alan Couch 57b; Jade Craven 73b; CSIRO Australian National Fish Collection 94b, 95t; CSIRO ScienceImage 51t, 64t,
72b, 76b; Ron DeCloux 24bl, 25b, 30t, 35t, 39tr, 115b, 116t&b; Jan Delsing 37t; Al Dobbin 34b, 44t; Patrick Doll 85b;
Ryan Dunn 97bl; Bernard DuPont 26t, 81b, 108b, 112t; Kane Durrant 67t; Graham Edgar 117t; Scott Eipper/Nature 4
You 44b, 47t, 48t&b, 49t,tr&b, 55b, 56, 59t, 63b, 112b, 122t&b, 123t&b, 124t&b, 125t&b, 126t&b, 127t,c&b, 128b,
129t&b, 130tr, 131t&b, 132t&b, 133t&b, 134t&b, 135t&b, 136trl,trr&b, 137tl,cl,cr&b, 138t&b, 139b, 140t&b, 141t&b,
142tl,trr&b, 143b, 144t&b, 145t&b, 146t, 147tl,tr&b, 148t&b, 149bl&br, 150t&b, 154b, 161t, 165b; Tie Eipper/Nature
4 You 15tl,tc,tr,bl&bc, 153t; Adam Elliott 45b, 47b, 69t; Damien Esquerre 53br, 57t, 84t, 98b, 100t, 104t, 106b, 108t,
118t; Ryan Francis 46t, 59b, 68b, 156t, 159t; Peter Fuller 110t; Dr. Lisa Gershwin 27b, 28b, 29t; Christian Gloor 121t;
Ratha Grimes 96b, 99t; David Guillemet 88b, 92b, 94t; Dr. Robert Hartwick/University of Hawaii 27t; Steve Herring
166t; Hans Hillewaert 32t; PT Hirschfield/Pink Tank Scuba (www.pinktankscuba.com) 33b, 109b, 113t, 115t; David
Hobern 75t, 77t&br, 78t&b; Anne Hoggett 29b; Nick Hopgood 25t; Brocken Inaglory 43b; Joi Ito 89b; Max Jackson
154t; Tony Jewel 73t; Albert Kok 90b, 91t; Kris-Mikael Krister 93b; Elias Levy 85t; Francois Libert 92t, 118b, 119t; Ross
McGibbon 139t; Sean McLean 79t; Angus McNab 130tl, 149t, 151t, 152t, 155, 158t&b, 159b, 160t, 161b, 162t&b,
163t, 165t, 166b; Mark Marathon 69b; Marinethemes.com/Kelvin Atkin 95b; Tam Warner Minton 90t; Susan Morrison
31b; National Oceanic and Atmospheric Administration 38tl, 93t, 114b; Katja Nevin 64b; James Niland 80t; Richard
Parker 36b, 37b, 38b, 39tl&b, 41b, 42t; Adam Parsons 52b, 53tl&tr, 54bl, 55tr, 58b; Karla Quintana Pearce 54br; Sue
& Rob Peatling 105t, 119b; Q. Phia 82t; Joe Pollock 24br; Sylke Rohrlach 42b, 87b, 96t, 98t, 102b, 103b, 104b, 107b,
109t, 113b, 120t; Mark Sanders/Ecosmart Ecology 164b; Udo Schmidt 36t, 40b, 41t; Katja Schulz 63t; Shawn Scott
130b, 146b; Licheng Shih 76t; Ian Skipworth 111t; Peter Southwood 28t, 32b, 83t; Peter Street 45tr, 67b, 74t&b, 75b,
77bl; Jason Sulda 58t, 110b; John Tann 79b; John Turnbull 30b, 33t, 35b, 82bl&br, 84b, 87t, 97t, 99b, 100b, 101t,
102t, 105b, 107t, 114t, 120b, 121b; Uncredited 38tr, 40t, 97br; US Department of Agriculture 60b; Alexander Vasenin
91b; Taso Viglas 83b, 86b; Nick Volpe 46b, 60t; Alex Wild 70t; Angell Williams 34t; Lou Wolfers 68tl&68tr; www.
kapeimages.com/Peter Rowland 15br, 31t, 45tl, 50t&b, 51b, 52t, 53bl, 54t, 55tl, 61tl,tr&b, 62t&b, 65bl&br, 66b, 70b,
71b, 71tl,tr&b, 156b, 157b, 160b, 163b, 164t; Anders Zimny 128t, 151b

ISBN 978-1-912081-60-8

Edited by Krystyna Mayer

Designed by Gulmohur Press, New Delhi

Printed and bound in Malaysia by Times Offset (M) Sdn. Bhd.

·CONTENTS·

INTRODUCTION

In 2016 a 22-year-old man died after being bitten by a Redback Spider during a bushwalk on the north coast of New South Wales. This was the first recorded fatality from a Redback Spider bite in 37 years (Redback Spider Antivenom was introduced in 1956). About 2,000 bites from this species are reported every year. The more deadly Sydney Funnel-web has been responsible for a further 30–40 bites each year, but no fatalities have been attributed to it since an antivenom to its bite was developed in 1981, and only 13 human deaths were recorded before then.

Snakes are responsible for more than 3,000 bites a year, but only an average of two of these prove fatal, and about 20 per cent of fatalities are a direct result of a person trying to handle or kill a snake. The venom of the box jellyfish *Chironex fleckeri* has also been attributed with almost 70 deaths. All venomous creatures are potentially deadly to humans, but how commonly an animal is encountered, the amount of venom it injects, the speed with which first aid and other medical treatment (including antivenom) are administered, and the age, health and sensitivity of the person envenomed are all contributing factors.

Since records began in 1791, there have been 1,062 documented cases of shark attacks on humans in Australia, with 237 resulting in fatalities, although 356 attacks (51 fatalities) were as the result of the shark being provoked. Since 1868, Saltwater Crocodiles have attacked and killed 77 people in Australia, from a total of 201 recorded incidents. Emus, kangaroos, cattle, horses and even camels all cause motor-vehicle crashes. Some of these animals also cause substantial damage to aircraft around airports. In the 10 years from 2006 to 2015, 101 collisions occurred with kangaroos and wallabies, 27 with dogs and foxes, and five with cattle. When birds are included in the statistics for the same 10-year period, the number of wildlife strikes on aircraft in Australia totals a staggering 16,069 – however, only 11 of these have resulted in serious damage to the aircraft and none resulted in fatalities.

While dangerous animal encounters are common, it is unusual for them to result in a serious injury and fatalities are rare. The aim of this book is not to leave the reader with a fear of Australian animals by over-emphasizing or over-dramatizing the potential they have to cause injury, illness or death to humans. Rather, it aims to raise awareness of the physical or chemical adaptations that the animals use to protect themselves, defend their territories or obtain the food they need for their survival and that of their young.

THE DIVERSITY OF AUSTRALIA'S DANGEROUS CREATURES

Australia is one of only 17 megadiverse countries in the world that collectively hold around two-thirds of the world's biodiversity. The number of vertebrate animal species in Australia is estimated at more than 10,000, and invertebrates are estimated at over 320,000.

INVERTEBRATES

Phylums (Porifera, Cnidaria, mollusca, Annelida, Arthropoda & Enchinodermata)
Invertebrates are animals that lack a backbone, and they make up around 90 per cent of known animal species. They can be loosely grouped as terrestrial invertebrates

and aquatic (both marine and freshwater) invertebrates, although many classes have members that occupy both groups and many terrestrial invertebrates have larval stages that are aquatic.

Sponges (Classes Calcarea, Demospongiae, Homoscleromorpha & Hexactinellida)
Sponges are marine animals that filter water through the small pores, or ostia, around the body, absorbing the nutrients they need to survive and expelling the waste through larger holes, or oscula. Some sponges are harmful to humans, as the toxins they absorb while feeding are stored and reused for metabolic purposes, including defence against predators and attacking neighbouring organisms in order to try and acquire more space in crowded environments.

Corals, Sea Anemones & Colonial Anemones (Class Anthozoa)
Members of this class of marine invertebrates are almost always encountered attached to a marine substrate on or close to the sea bottom. The group includes the hard and soft corals, sea pens, sea fans, gorgonians and sea anemones, some of which are poisonous. Palytoxin was first identified in the zoanthid coral *Palythoa toxica*, a coral found in the waters around Hawaii. Palytoxin is regarded as the second most toxic naturally occurring non-protein substance in the world. It is also found in other members of the *Palythoa* genus, and certain dinoflagellates.

Box Jellyfish (Class Cubozoa)
This group of cnidarians is characterized by the cube-shaped bell (medusa) possessed by its members. They are fast-swimming invertebrates that have well-developed eyes for locating prey, mainly zooplankton, crustaceans and fish, which is trapped and immobilized by the venomous tentacles and moved to the mouth, usually while inverted. Although some cubozoans have been responsible for many human deaths, not all species possess venom of this potency. The box jellyfish *Chironex fleckeri* has been responsible for around 70 human deaths. The much smaller Irukandji jellyfish (family Carukiidae) causes collective symptoms in humans known as 'Irukandji Syndrome', which manifests as a sharp prickling sensation, followed sometime after by lower back pain, muscle cramping, respiratory distress, headache, nausea, vomiting and depression. Similar symptoms are produced by other box jellyfish species and are referred to as 'Irukandji-like Syndrome'.

Hydrozoans (Class Hydrozoa)
This diverse group of cnidarians includes jellyfish, corals, hydras, sea anemones, sea ferns and related organisms. They inhabit both marine and freshwater environments, and can be colonial (mainly) or solitary in nature. Hydrozoans are mostly characterized by a life cycle that contains a planula larval stage (which normally develops into a sessile polyp) and the medusa stage, a free-swimming breeding stage. They all contain stinging cells called nematocysts, or cnida, which are used to capture prey or repel predators, and some can cause severe reactions in humans.

True Jellyfish (**Class** Scyphozoa)

Scyphozoans are widespread throughout the world's oceans, from the colder waters of the Arctic and Antarctic to the tropics, where they propel themselves through the water by contracting and relaxing the muscles of their bell. Most live in shallow water around the coasts, but some inhabit deeper oceans. Like the other three classes that make up the phylum Cnidaria (the anthozoans, cubozoans and hydrozoans), the scyphozoans have stinging cells, called nematocysts, which are used for hunting prey and defence against predators. The stinging cells of some species are capable of causing intense pain in humans, and allergic reactions in people with sensitivity to the venom.

Octopuses, Squids & Cuttlefish (**Class** Cephalopoda)

There are currently more than 800 species in this mollusc group, which are widespread in the world's oceans. They are reputed to have a high intelligence and are famed for their incredible camouflaging skills, being able to change colour, pattern and shape. The blue-ringed and blue-lined octopuses have been responsible for a number of human deaths, although bites generally only occur if individuals are removed from the water and placed on the skin. The first recorded fatality is attributed to the Northern Australian Greater Blue-ringed Octopus. The octopuses bite through the skin and inject a toxic saliva. Some other members of these groups have toxic body tissues.

Snails & Slugs (**Class** Gastropoda)

This very large group of diverse molluscs is widespread in both terrestrial and aquatic environments, and each species has a characteristic single muscular foot that is used for locomotion. The snails have a coiled shell into which the body can be withdrawn in most species (the exception being the semi-slugs), but this feature has been reduced to a small internal remnant or lost completely in the slugs. The nudibranchs, or sea slugs, have larvae with a coiled shell, but this is lost when they metamorphose into the adult form.

Cone snails are carnivorous gastropods, actively hunting fish, worms and other molluscs (mainly other snails), spearing their victims with a highly modified radular tooth and injecting them with venom that causes rapid paralysis of the muscle tissues. They are mostly nocturnal but are still able to inflict a deadly sting when at rest during the day. There is no antivenom for cone snail venom. The beautiful colouration and patterning of the shells makes them attractive to shell collectors and traders, and as playthings for children.

The sea slugs of the genus *Glaucus* feed on jellyfish, especially Portuguese Man O'Wars, and store the nematocysts of their victims in the tips of their cerata for their own protection, giving them the potential to have a high concentration of venom.

Bristleworms (**Class** Polychaeta)

These segmented invertebrates usually have a well-defined head and leg-like appendages, or parapodia, on every body segment, each of which is covered with finer bristles, or chaetae. The chaetae contain a toxin that can cause intense pain and severe allergic reactions, including anaphylaxis. Bristleworms are among the most common marine organisms, and can be found in a variety of marine habitats and at a range of depths.

Leeches (Class Clitellata)

These segmented invertebrates usually have a poorly defined head and lack parapodia ('legs'). The mouth is on the undersurface and the brain is located in one of the body segments. The jawless leeches are known to harbour malarial parasites, but are not known to transmit these to humans. While there is no scientific evidence that the jawed leeches that feed on human blood transmit diseases to their hosts, allergic reactions are not uncommon and the wound can become infected. It can also bleed and stay itchy and inflamed for several hours after the leech has dropped off.

Most leeches can be removed safely if discovered feeding on the human body. They can be made to drop off by touching them with a hot object, salt, eucalyptus oil or tea tree oil. However, they have been known to attach themselves to an eyeball, when they should not be removed forcibly. Any bite site should be treated with antiseptic to prevent infections, particularly if the leeches have been pulled off by hand.

Leeches are also capable of causing pruritus and palpable purpura. They have been used for medicinal purposes for many years, and are utilized today to assist in the reattachment of severed body parts, and in plastic surgery to reduce blood clotting in the finer tissues.

Spiders, Scorpions & Allies (Class Arachnida)

Australia has a reputation for having more than its fair share of dangerous spiders, with the funnel-webs and Redback Spider topping the list. Spiders tend to instill more fear in humans than many other animal groups, but deaths from their bites are very rare, with most bites resulting in localized pain and swelling. Yearly, around 2,000 people seek medical attention for Redback Spider bites, and up to 40 for bites from funnel-web spiders.

At least 13 human deaths have been caused by the 40-plus species of funnel-web spider known to science. The Sydney Funnel-web is attributed with most of these, and is widely regarded as Australia's most deadly spider species and among the most deadly spiders in the world. Since the development of an antivenom in 1981, no human deaths from funnel-web bites have been recorded. Not all bites result in sufficient quantities of venom being injected to produce severe envenomation, but all bites should be treated as potentially fatal and urgent medical attention should be sought. Another family of spiders with a fearsome reputation is the white-tailed spiders. The bites of some species in this family can be very painful in humans and some result in tissue necrosis (necrotizing arachnidism), which most people incorrectly believe to be caused by the venom, but is actually caused by infection of the bite site with the bacterium *Mycobacterium ulcerans*.

Scorpions are also venomous, but none of the species in Australia is generally considered lethal. There have been two cases of fatal stings in Australia, both in infants. One was in Western Australia in 1929 by *Lychas marmoreus*, and the other in Tasmania in 1932 by *Cercophonius squama*. The sting can be very painful, with an amount of swelling and itchiness also evident, but symptoms normally pass after several hours. Nonetheless, medical advice should always be sought if a person is stung by a scorpion, as severe reactions, including anaphylaxis, can occur in some people.

Other members of this family include ticks and mites. These small arachnids have a typically broad, oval body and small cephalothorax. They may be terrestrial or aquatic, and feed on either plants or animals. Most of the more than 30,000 species are free living, but some are parasitic and capable of spreading disease organisms, including spirochete bacteria, which cause Lyme disease and relapsing fever. Some victims can develop a syndrome known as 'tick-induced mammalian meat allergy', which results in a sensitivity to meat products such as gelatin, causing anaphylaxis. Queensland Tick Typhus and Flinders Island Spotted Fever are other known diseases caused by tick bites.

Centipedes (Class Chilopoda)

These predatory terrestrial invertebrates have up to 177 flattened body segments, most of which have a single pair of legs. The first body segment has a pair of large venomous fangs, or forcipules, which are used for hunting and are a modified pair of legs. Centipedes can give an extremely painful bite, although usually without long-lasting systemic effects.

Insects (Class Insecta)

Insects comprise an incredibly diverse assemblage of animals and are the most numerous multicellular organisms on the planet, with more than one million species currently identified. They are invertebrates (they do not have a backbone), and can be found in almost every habitat worldwide.

The Honey Bee, including the hybrid form known as the Africanized Honey Bee, or killer bee, is one of the most dangerous animals on Earth, due to the severe allergic reactions (anaphylaxis) it produces in people with a sensitivity to its venom. On average about 1,000 people are hospitalized due to allergic reactions to bee venom each year in Australia. While only a very small percentage of them die from the sting, around half of the deaths occur before reaching hospital.. People with a known allergy to bee stings carry adrenalin pens in case they are stung. Wasps and ants are closely related to the Honey Bee, and differ from it in that their stingers are not barbed, so they are able to sting their prey or attack multiple times in quick succession. Their stings can also cause severe pain, burning and itching, and like the Honey Bee's sting can produce life-threatening allergic reactions in some people.

Mosquitoes are well known for spreading diseases, with malaria being perhaps the most infamous disease. Malaria is caused by a group of single-celled parasites that belong to the genus *Plasmodium*. Australia has been free from malaria since 1981, but people that have been infected while travelling overseas, including to Papua New Guinea – Australia's closest neighbour and home to the deadly cerebral malaria protozoa *Plasmodium falciparum*, which attacks the human brain – have been diagnosed after they arrive in Australia. Within Australia, the main viruses that mosquitoes are responsible for transmitting are Ross River virus, Dengue fever, Murray River encephalitis virus, Kunjin virus and Barmah Forest virus, all of which belong to a group known as arboviruses.

Dengue fever is the most significant arbovirus worldwide, and around 1,000 reported cases occur in Far North Queensland each year. There are varying types of the virus, called serotypes, and symptoms caused by these can range from mild fevers, headaches and joint and muscle pain, to internal bleeding (Dengue haemorrhagic fever) and shock (Dengue shock syndrome), both of which are potentially fatal. The only confirmed vector for Dengue fever in Australia is the Yellow Fever Mosquito *Aedes aegypti*, and transmission of the virus only occurs from an infected human.

There are more than 40 known vector mosquitoes for Ross River virus, and 12 are known for Barmah Forest virus. Ross River virus was first isolated from the Saltmarsh Mosquito A. *vigilax* in 1959, and Barmah River virus from the Common Banded Mosquito *Culex annulirostris* in 1974. Both these viruses cause joint and muscle pain, rashes, fatigue, headaches and swollen lymph nodes, and about 6,000 people, from all states and territories in Australia, are infected each year. Murray Valley encephalitis is caused by the Murray Valley encephalitis virus and Kunjin virus, which uses waterbirds as the primary host and causes varying degrees of neurological impairment. These viruses can be permanent and potentially fatal when transmitted to humans, although it is thought that only one in 1,000 people develops symptoms.

Some species of fly, like the House Fly and Australian Bush Fly, carry microscopic organisms on their hairy bodies that can transmit diseases by coming into contact with foodstuffs and objects that are used by humans. Other species, like the Stable Fly, are active feeders on blood, readily attacking humans, pets and commercial livestock. They also feed on composting material. Although they are capable of spreading pathogens and diseases, including typhoid fever, dysentery and anthrax, this is not thought to occur in Australia, though their bites can be very painful. March Flies are capable of spreading disease and parasites to humans, including loiasis, tularemia and anthrax, although this is not thought to occur in Australia, but the bite can lead to allergic reactions in some people. Widespread throughout warm regions of Australia, they tend to be most active during the warmest months.

After biting, female biting midges (known by most Australians as sandflies) inject an anticoagulant to prevent the blood from clotting, which can cause a reaction in people such as swelling, itchiness, blisters and sores.

Some species of flea, such as the Cat Flea *Ctenocephalides felis*, feed on a range of hosts, including cats, dogs and humans. This species has the potential to transfer diseases from one host to another. It is also a known host for cat tapeworm. The Rat Flea is known to cause murine typhus in humans in Australia – this results in chills, headache, fever, rashes and potentially death. Other species known to commonly attack humans are the Dog Flea C. *canis* and the Human Flea *Pulex irritans*, but both are less common. Flea bites can also cause lesions, itchiness and secondary infection.

Lice are responsible for the transmission of disease. There are two main suborders of louse: Anoplura, which feeds mainly on mammals, and Mallophaga, which feeds mainly on both birds and mammals. There are more than 3,000 lice species worldwide, around 200 of which occur in Australia. Three types of louse are known to live on humans: the Head Louse *Pediculus humanus capitas*, Human Body Louse *P. h. corporis* and Pubic Louse (often referred to as crab louse) *Pthirus pubis*, but only the body louse is known to spread diseases.

Sea Stars (Class Asteroidea)

Distributed widely throughout the world in marine and brackish waters, the sea stars, or starfish, mainly prey on sponges, algal slime and molluscs, which are either hunted or scavenged along the bottom of their aquatic home. The Crown-of-thorns Starfish has spines on the upper surface that are regarded as venomous, causing intense pain, swelling, redness, vomiting and potential anaphylaxis in humans, and it has been responsible for at least one human fatality.

Echinoids (Class Echinoidea)

Echinoids, including sea urchins, sand dollars, pencil urchins, heart urchins and related organisms, are mostly characterized by tightly interlocking plates that form the rigid or slightly flexible exoskeleton, or test. Some sea urchins are known to have venomous spines. Both the Black Longspine Urchin and Black-spined Sea Urchin have brittle spines that contain venom. The spines easily pierce human skin and break off, producing pain, swelling and a potential source of secondary infection. The Flower Urchin has pincer-like pedicellariae that have fang-like tips, which inject venom that causes intense pain, paralysis and respiratory distress, and is known to have caused fatalities.

VERTEBRATES (Phylum Chordata)

Vertebrates consist of five or seven classes, depending on the authority. They are mammals, birds, reptiles, amphibians, jawless fish, cartilaginous fish and bony fish (the latter three sometimes combined into a single class), each characterized by the presence of a notochord or backbone (vertebral column).

Cartilaginous Fish (Class Chondrichthyes)

This is the oldest group of modern-day jawed vertebrates, which diverged from common shark-like ancestors around 250 million years ago (mya), with some members possibly living over 400 mya. All members of the group have a cartilaginous skeleton.

Both sharks and rays have been responsible for human deaths. Over the past 100 years, there has been a total of 869 reported shark attacks on humans, 171 of which have proven fatal and a further 171 of which did not result in any injury to the people attacked. Out of this total, 567 attacks, including 134 fatalities, were unprovoked.

Both the Bull Shark and Tiger Shark belong to the family Charcharinidae (whaler sharks), and most attacks on humans are attributed to members of this family, although similarities between several species make positive identification of the actual species responsible for many attacks difficult. Some whaler sharks can be territorial and more aggressive at certain times, possibly when breeding, and they usually give warnings involving exaggerated swimming movements with an arched back before an attack is made. People should exit the water as soon as possible when this behaviour occurs. Although docile, species from the carpet shark and wobbegong family have been listed as responsible for 51 reported shark attack cases between 1900 and 2009, with 24 of the attacks occurring in the last 10 years of that period. No fatalities have been recorded, with most attacks being regarded as a defensive reaction after people stepped on the sharks or got too close to them.

Many rays are equipped with venomous barbs that are used for defence against aggressors. Due to their mostly bottom-dwelling habits, unwitting people are likely to stand on them when walking in shallow water, and the barbs are capable of causing intense pain.

The chimaeras possess venomous spines that can inflict a painful wound in humans, but do not appear to pose a major danger.

Bony Fish (Class Actinopterygii)

The ray-finned fish *Actinopterygii* are the largest assemblage of all vertebrate groups, with more than 29,000 species known. Together with the *Sarcopterygii*, the lobe-finned fish and tetrapods, they are the bony fish, or osteichthyes. Ray-finned fish are so named due to the membranes of skin (fins) on the body, which are supported by stiff spines (rays). In some species the spines are capable of injecting toxic venom, which is used as a form of defence against any potential predators that try to eat them (or humans that try to handle them).

There are numerous venomous bony fish species in Australia's waters, and the venom contained in glands at the base of the dorsal spines of the two stonefish species is believed to be the most potent of all of the world's fish. It causes extremely painful wounds when injected and can prove fatal in some situations; an antivenom was developed in 1959. Some members of the frogfish family have venomous spines, and some Australian species are thought to also contain venom. Although not normally encountered by most people, members of several fish families, such as the scorpionfishes, have a similar armoury of venomous spines. One family that is regularly encountered, however, is the flatheads, which are a popular target for commercial and recreational fishers as they are regarded as an excellent eating fish, with many tonnes caught in all Australian states every year. They possess venomous opercular spines, and care needs to be taken when handling them.

Ciguatera is a form of fish poisoning caused by the ingestion by herbivorous fish of the small marine organism *Gambierdiscus toxicus*, which grows on coral reefs. After another fish eats the herbivorous fish, the toxins it contains are modified and more concentrated as they move up the food chain. Consumption of large carnivorous reef fish, including moray eels, barracudas, Chinaman, red sea bass, Spanish mackerel and trevally, can cause ciguatera poisoning in humans, generally from fish in tropical waters down to northern New South Wales in autumn. Symptoms include diarrhoea, nausea, vomiting, heart problems, nerve damage, insomnia and fatigue, and potentially death (depending on the amount of toxin ingested). Moray eels, particularly of the genus *Gymnothorax*, should never be eaten, especially if they are ungutted, and even small fish have been known to cause human fatalities. The sale of moray eels as food is banned in Australia.

Some pufferfish species are thought to be the second most toxic vertebrates in the world after the Golden Poison Frog *Phyllobates terribilis* of Colombia. Their internal organs contain a deadly poison (tetrodotoxin), which is most concentrated in the liver, ovaries and skin, and when consumed in sufficient amounts can kill a human. The toxin is water soluble, and cooking increases its toxic effect. In Japan and Korea pufferfish species are eaten (known respectively as Fugu and Bogeo), but the food preparation is undertaken by trained professional chefs and strictly controlled by law.

Another poison, known as pahutoxin, a type of neurotoxin, is secreted by many (possibly all) boxfish members. The toxin is secreted by the animal from its skin when stressed, and some species advertise this to would-be attackers with their bright colouration.

Amphibians (Class Amphibia)

Amphibians were the first vertebrates to colonize the land more than 370 mya, and most still require water for the larval stage of their life cycle. The group includes the salamanders, frogs, toads and caecilians, all of which have a three-chambered heart. Some Australian frogs produce alkaloid toxins that are similar to that of the poison dart frogs of South America, which are thought to be harmful to mammals.

Reptiles (Class Reptilia)

The reptiles are a diverse group of vertebrate animals that includes lizards, snakes, tuataras, turtles, amphisbaenians and crocodiles, all of which have scales. They are ectothermic, with their body temperature varying depending on their surroundings, and many species can be seen basking in sunlight or absorbing heat from rocks, and also on roads, to enable them to perform daily activities such as hunting or digesting food.

Australia's largest terrestrial carnivore is the Saltwater Crocodile, which has been responsible for 74 human deaths in the tropical north since 1869, with 17 occurring in the past 10 years, although there have been about four times as many attacks. The other crocodile species in Australia, the Freshwater Crocodile, is capable of inflicting painful bites that can easily become infected, but it is generally regarded as inoffensive, only biting people that have disturbed it in some way. There are also massive individuals among the pythons. The largest of the python species, such as the Scrub Python, are capable of overpowering an adult human and causing death from constriction.

Australia is well known for its venomous snakes, and while the venom of some species produces only mild effects on humans, other species, such as the Coastal Taipan, Eastern Brown Snake and Tiger Snake, have extremely toxic venom and have been responsible for human fatalities. However, most bites on humans do not result in the snake injecting a sufficient amount of venom to cause life-threatening side-effects, serving more as a warning to the aggressor, although all bites should be treated as potentially serious. Snakes are not naturally aggressive towards humans, and most instinctively move away from potential danger if disturbed.

Australian varanids (monitor lizards, or goannas) have sharp claws and powerful limbs and jaws, which can inflict painful injuries if the animals are handled. Recent studies have indicated that Australian varanids may also produce venom, though its purpose is uncertain.

Birds (Class Aves)

Birds are probably the most familiar vertebrate animals around the world, occupying every continent and every habitat type. They are warm blooded, or endothermic, have feathers and lay eggs that are protected by hard outer shells. The heaviest species, the Southern Cassowary, has been responsible for fatally wounding a 16-year-old boy after it was provoked. The biggest threat that birds pose to humans lies in the damage they cause to aircraft – they have been responsible for more than 16,000 strikes on aircraft between 2006 and 2015.

Mammals (Class Mammalia)

About 386 species of mammal are found in Australia, around 7 per cent of the world's total, and 87 per cent of these are endemic to Australia and 20 per cent are considered threatened. There are three main mammal groups: monotremes, marsupials and placentals.

The Platypus is one of about 13 venomous mammals in the world and Australia's only representative. The male produces a powerful venom during the breeding season, which it injects into its rivals during competition for females and while defending its territory.

Australia also has the Dingo, an introduced dog that shares a common ancestral path with the Domestic Dog. Both have been known to attack humans, and some of the latter have killed humans. This has led to certain dog breeds being restricted in Australia, including Argentinian and Brazilian Mastiffs. For the most part dogs only present a risk when they perceive a threat or if they have been abused. Good training from birth with appropriate supervision for the dogs, along with teaching basic dog behaviour to school-age children, would significantly mitigate the risk dogs pose to people.

Flying-foxes, or fruit-bats, are responsible for more strikes on aircraft than any other flying animal in Australia, with the majority of strikes occurring on the east coast. One member of the smaller insectivorous bats, the Yellow-bellied Sheathtail Bat, has been identified as a carrier of the deadly Australian Bat Lyssavirus (ABLV). This is one of 12 types of *Lyssavirus* that have been recorded worldwide. It can be prevented by the rabies vaccine, but once the disease manifests itself in a human there is no current cure. There have been three recorded cases of ABLV in Australia, all of which have been fatal.

AVOIDING DANGEROUS ENCOUNTERS

CROCODILES

- Do not enter water that may contain crocodiles (often indicated by signs, but not always, so check with local authorities).
- Do not clean fish on the edge of a waterway in areas where crocodiles are present.
- Do not camp next to a waterway.
- Be aware that crocodiles can be present in the ocean as well as in freshwater areas.
- Respect crocodiles and view them from afar – they can be extremely fast moving and are often underestimated.

SNAKES & ARTHROPODS

- Never attempt to catch a snake.
- Never attempt to kill a snake – this is not only illegal, but will also put you at serious risk.
- Never place your hands and feet where you cannot see them.
- Never walk around at night without a torch.
- Always wear shoes and long trousers in locations where these creatures are likely to be present.
- Keep rodents away from houses.
- Keep gardens and sheds tidy and free of clutter.
- Keep birds away from houses – their food attracts rodents, which in turn attract snakes.
- Take care when turning over potential shelter sites such as corrugated iron sheets, rocks and logs, as well as other debris.
- Store shoes indoors so nothing can seek out shelter inside a shoe.
- Stay away from nests and hives.
- Use approved insect repellents and accessories (such as face and head nets).
- Install and maintain screens on windows and doors.

VENOMOUS AQUATIC SPECIES

- Wear good shoes when walking through rivers and estuaries, on reefs and in intertidal zones.
- Do not handle or otherwise provoke wildlife, and avoid handling shells that might still contain an animal.
- Wear gloves when handling fish.
- Take care when removing fish and other creatures from nets and lines.
- Do not antagonize fish when diving.
- Always wear a stinger suit when diving or swimming in northern Australia.

FIRST AID

DIAL '000' (TRIPLE ZERO) FOR AN AMBULANCE IN ANY MEDICAL EMERGENCY.

If someone is bitten or stung by an animal, the early application of the correct first-aid practices can greatly increase their chances of making a full recovery. First-aid measures are constantly advancing and improving, so it is important to stay current with the latest practices.

In the first instance

The DRSABCD action plan should be followed:

D look for **D**anger
R check for **R**esponse
S **S**end for help
A clear the **A**irway
B sustain **B**reathing
C start **C**PR (if required)
D apply a **D**efibrillator if indicated

DRSABCD is vital if a person has collapsed and is unresponsive.

Snakebite

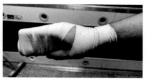

Bandage over wound from end of limb

Bandage complete limb

Immoblized limb

Mark bite site and time

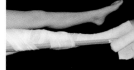

Bite to lower limb showing splint

PRESSURE IMMOBILIZATION BANDAGING (PIB) FIRST AID

This is currently the recommended treatment for bites and stings of Australian venomous snakes, funnel-web spiders, Blue-ringed Octopuses and cone shells.

There are two components that must be satisfied – pressure over the bitten limb and focal plus general immobilization. This involves the application of:

1. A broad (minimum 75mm-wide) elastic bandage to the entire bitten limb at a very firm pressure of at least 40mmHg for an arm and 55mmHg for a leg. The Australian Venom Research Unit (AVRU) recommends SETOPRESS TM High Compression Bandages as these bandages relax very little with prolonged application.

2. Splints to effectivity immobilize the entire limb, in combination with laying the person down and ensuring that they are completely still to minimize any movement. Do not use a sling.

Move away from the area where the bite or sting occurred (if required), lay the person down and keep them calm. Any movement of the limb quickly results in venom absorption and must be prevented; therefore first aid must be an immediate priority after a snakebite.

Do not allow the person to walk. The recommendation is that in the case of a snakebite to a lower limb, splinting of both legs should be carried out to completely immobilize the lower half of the body.

In rare cases a person may be bitten on the body, face or neck. In these cases direct pressure should be applied over the bite site with a pressure pad made from a cloth (such as a hand towel or T-shirt) and held firmly in place until medical attention can be sought. Always seek medical attention following a snakebite.

VENOMOUS SNAKE, FUNNEL-WEB SPIDER, MOUSE SPIDER & BLUE-RINGED OCTOPUS BITES, & CONE SNAIL STINGS

- *Seek urgent medical help by dialing triple zero* '000'.
- *Do not* wash the bite site.
- Apply a pressure immobilization bandage and splint (if a limb).
- Monitor the person's breathing, pulse and circulation to the extremities, and begin CPR (30 chest compressions followed by two rescue breaths) if required, and continue until the person's breathing is normal and stable.
- *Do not* catch, chase or kill the snake, spider, octopus or cone snail – this involves extra movement (if done by the person affected), and could result in further bites or stings.
- *Do not* drink alcohol or tea, or take stimulants, food or medication without expert advice.
- *Do not* wash the wound, or apply hot or cold packs, electrical shocks, suction devices or tourniquets/ligatures.

REDBACK SPIDER BITES

- Wash the bite site.
- Apply ice packs.
- *Do not* apply a bandage.
- Seek medical attention.

TICK BITES

- *Do not* use methylated spirits or alcohol to kill a tick before removal.
- Remove the tick using fine tweezers, by grasping it behind the head (as close to the affected person's skin as possible), and gently pull straight (keep the tick for identification).
- Ensure that all parts of the tick have been removed.
- *Do not* squeeze the tick's body.
- Wash the site with warm water.
- Apply an antiseptic.
- Apply a local bandage (such as a band aid) for 24 hours.
- Seek medical attention if more severe symptoms develop.

BEE STINGS

- Remove the stinger barb by scraping sideways.
- *Do not* squeeze the venom sac dislodged from the bee's abdomen.
- Wash the site.
- Apply ice packs.
- *If the person is allergic:*
  - Remove the stinger barb, if still present.
  - Lay the person flat, or put them in a sitting position if breathing is difficult.
  - Follow the guidelines under 'Management of Difficulty in Breathing' (see p. 18).
  - Dial triple zero '000'.
  - *Do not* allow the person to move around.

WASP STINGS

- Wash the site with warm, soapy water.
- Apply ice packs.
- *If the person is allergic:*
  - Remove the stinger, if still present.
  - Lay the person flat, or put them in a sitting position if breathing is difficult.
  - Follow the guidelines under 'Management of Difficulty in Breathing' (see p. 18).
  - Dial triple zero '000'.
  - *Do not* allow the person to move around.

STONEFISH, STINGRAYS & OTHER VENOMOUS FISH STINGS

- *Do not* remove any penetrating barbs (stingrays).
- Any stingray wound to the trunk of the body should be treated as a medical emergency – seek urgent medical attention by dialing triple zero '000'.
- Stem any major bleeding as the highest priority.
- Immerse the site in hot water (45° C), *not* scalding water, for 20 minutes.
- Remove briefly and repeat the process until the pain subsides (*not* for longer than two hours).
- If symptoms persist, seek urgent medical attention by dialing triple zero '000' (*antivenom is available for stonefish*).
- Any lacerated wounds should be checked and treated by a medical professional for secondary infections.
- *If the person is allergic:*
  - Remove the allergen, if still present.
  - Lay the person flat, or put them in a sitting position if breathing is difficult.
  - Follow the guidelines under 'Management of Difficulty in Breathing' (see p. 18).
  - Dial triple zero '000'.
  - *Do not* allow the person to move around.

SEA URCHIN & CROWN-OF-THORNS STARFISH STINGS

- Immerse the site in hot water (45° C), *not* scalding water, for 20 minutes.
- Remove briefly and repeat the process until the pain subsides (*not* longer than two hours).
- If symptoms persist or worsen, seek urgent medical attention by dialing triple zero **'000'**.
- *If the person is allergic:*
  - Remove the allergen, if still present.
  - Lay the person flat, or or put them in a sitting position if breathing is difficult.
  - Follow the guidelines under 'Management of Difficulty in Breathing' (see below).
  - Dial triple zero **'000'**.
  - *Do not* allow the person to move around.

CNIDARIAN STINGS (INCLUDING BOX JELLYFISH, TRUE JELLYFISH & PORTUGUESE MAN O'WARS)

Continuing studies and clinical trials are constantly reviewing and updating these procedures, and readers are recommended to remain up to date with current first-aid guidelines from scientific trials. First-aid measures may change for individual species.

- *Seek urgent medical help by dialing triple zero* **'000'**.
- *Do not* scrape the sting site to remove tentacles and stinging cells. Evidence suggests that less than 1 per cent of stinging cells discharge on initial contact, so poor removal techniques can dramatically increase the amount of venom injected into a person.
- *Do not* wash the area with sea water *or* scrub it with sand.
- Flood the site generously with vinegar for at least 30 seconds.
- If vinegar is not available, *carefully* remove the tentacles by 'plucking'.
- Immerse the area in hot water (45° C) for 45 minutes.
- *Do not* apply ice packs.
- *Do not* use ethanol, which increases pain and the discharge of stinging cells.
- Monitor the person's breathing, pulse and circulation to the extremities.

MANAGEMENT OF DIFFICULTY IN BREATHING

Envenomation-related difficulty in breathing can be due to true anaphylaxis as well as to anaphylactoid reactions.

Anaphylaxis is an IgE (immunoglobulin E) mediated response and requires previous exposure. Symptoms include acute, rapid-onset illness, with typical clinical symptoms appearing as:

- Tingling around the mouth.
- Swelling of the lips, tongue and face.
- Tightness in the throat.
- Difficult or noisy breathing.
- Difficulty talking.

- Coughing.
- Dizziness.
- Vomiting or abdominal pain, particularly following insect stings.
- Paleness and floppiness (in infants and young children).

If anaphylaxis is suspected:
- *Seek urgent medical help by dialing triple zero* **'000'**.
- Lay the person flat, or put them in a seated position if breathing is difficult.
- Obtain and follow the instructions on an epinephrine (adrenaline) pen – usually injected in the outer thigh.
- *Do not* allow the person to stand or walk around.
- Monitor the breathing and commence CPR (30 chest compressions followed by two rescue breaths) if required, and continue until the person's breathing is normal and stable.
- Monitor the person for at least four hours.

Anaphylactoid reactions produce the same clinical picture as does anaphylaxis, but are not IgE mediated. They may occur through a direct non-immune-mediated release of mediators by venom toxins from mast cells and/or basophils, or result from direct complement activation by venom components. *Only true IgE mediated Analyphylaxis should be managed by the administration of epinephrine.*
- Box jelly stings, as well as certain other jelly stings, can lead to respiratory insufficiency but in these cases 'catecholamine surge' is possible, in which the person undergoes a hypertensive phase. *Administration of epinephrine for anaphylactoid reactions due to jellyfish stings can lead to end organ failure and cerebral haemorrhage.*

THE SPECIES ACCOUNTS

A standard structure has been used in the species accounts. Abbreviations and features are explained below.

DISTRIBUTION KEY

NSW	New South Wales
Vic	Victoria
Tas	Tasmania
SA	South Australia
WA	Western Australia
NT	Northern Territory
Qld	Queensland
PNG	Papua New Guinea
NZ	New Zealand
GBR	Great Barrier Reef

DANGEROUS CREATURE RANKINGS

DANGEROUSLY VENOMOUS Venom can kill a healthy person (adult or child).
DANGEROUS Able to kill from mechanical injuries caused by teeth, claws or crushing.
Poisonous Contains toxins that can result in sickness and potential death if consumed, inhaled or absorbed through the skin in sufficient quantities.
Venomous Venom unlikely to kill a healthy person.
Potentially Dangerous Could kill a person in unusual circumstances or spread disease.
Harmful Non-lethal venom, contains irritants, is usually docile or carries low risk of disease, and is generally inoffensive, except in people who are at risk of severe allergic reactions and anaphylaxis.

KEY FEATURES & MEASUREMENTS

Sizes quoted in species accounts for body measurements (where available) are average maximum sizes, but exceptions can occur.

BH	Bell height
BW	Bell width
CL	Carapace length
DW	Disc width
H	Height
TL	Total length
W	Width

Glossary

abdomen Rear segment of body, behind thorax.
anaphylaxis Severe fast-acting allergic reaction that can be fatal.
anterior Front of body.
anticoagulant Substance that prevents clotting of blood.
antivenom Medication used to treat venomous bites and stings.
aquatic Living in fresh or salt water.
arboreal Living in or climbing trees for food or shelter.
asexual In reproduction, female able to reproduce without male, and offspring are clones.
basking Act of an animal exposing itself to increased temperature in order to raise its core body temperature.
barbels (on fish) Sensory, whisker-like organs around mouth.
benthic Refers to lowest ecological zone in body of water.
brackish water Water that is slightly salty or briny.
canines Pointed teeth between incisors and premolars.
carapace Hardened upper shell, as in turtles.
carcass Dead body of an animal.
cardiotoxic Denotes components of venom that attack cells of heart.

carnivorous Feeding only on animal matter.

carrion Flesh of dead animals.

cathemeral Active at any time of day or night.

caudal At or near back half of body.

cephalopod Member of the class Cephalopoda, which encompasses squid, octopuses and nautiluses.

cephalothorax Head and thorax of various arthropods, as in spiders.

cerata (*singular* **ceras**) Blood-filled tubular outgrowths.

cnidosacs Internal sac of nudibranchs that contains nematocysts recycled from their prey.

coagulant Component of venom that causes the blood to clot.

constriction Act of coiling tightly around an animal, causing suffocation.

crepuscular Active at dawn and dusk.

cuticle Protective outer layer of a hair shaft

cytotoxic Denotes components of venom that attack cells of body.

digit Toe or finger.

dimorphism Distinct physical differences between sexes.

distal Section of limb or attachment furthest from body.

diurnal Active during day.

dorsal Upper surface, or back.

ectoderm External tissue.

electrocyte Modified muscle cell capable of delivering an electrical discharge.

endemic Found only in a certain area.

envenom Inject venom into.

exoskeleton External skeleton used as protection, as in insects.

family Taxonomic rank above genus and below order.

fangs Set of sharp teeth for tearing flesh (mammals), or modified to pierce living tissue and inject venom (snakes and spiders).

feral Of an introduced animal that has become established in the wild.

fin Appendage, typically on fish, that helps control movement in water.

fossorial Living or active beneath soil surface; burrowing.

genus (*plural* **genera**) Taxonomic group above species and below family.

gill opening (slit) Single (bony fish) or multiple (cartilaginous fish) external openings behind head that lead to gill chamber.

gravid Pregnant; carrying eggs or young.

gregarious Living within group or community.

haemolytic/haemolysin Denotes components of venom that cause red blood cells to be destroyed.

herbivorous Feeding only on plant matter.

hermaphrodite Animal possessing both male and female sex organs.

herpetofauna Collective term referring to group of amphibians and reptiles – usually pertaining to particular region, locality or habitat.

heterocercal (of a shark's tail) With long upper lobe and shorter lower lobe, which assists in keeping snout in downwards direction.

incisors Chisel-shaped front teeth used for biting or gnawing.

infraorder Taxonomic rank below suborder.

insectivorous Feeding solely on insects.

instar Developmental stage in insect larva between moulting periods.

invertebrate Animal that lacks backbone.

iridophores Components within skin that form iridescent or reflective sheen in certain cephalopods.

larva (*plural* **larvae**) Newly hatched, wingless stage of an animal.

lateral Side of an animal.

mandible Jaw (lower).

marine Living in or part of aquatic environments where water salinity is greater than 30g of salt per 1kg of water.

maxilla Jaw (upper).

membrane Thin layer of tissue.

mob Small social group (specifically of macropods).

molars Rear teeth use to chew or grind.

morphological Pertaining to external appearance of an animal.

myotoxic Denotes components of venom that attack muscle tissue.

nape Back of neck.

necrosis Premature death of cells in living tissue.

nectarivorous Feeding purely on nectar.

nematocyst Specialized cell that projects harpoon-like thread in order to envenomate prey or attackers.

neonate Newborn animal.

neurotoxin Denotes components of venom that attack nervous system.

nocturnal Active at night.

nuchal In neck area.

omnivorous Feeding on both animal and plant matter.

order Taxonomic rank above family and below class.

oviparous Reproducing by laying eggs.

ovoviviparous Reproducing by forming unshelled egg sacs to house developing young, which are held inside female until ready to hatch, then expelled either still within egg sac or after they leave it.

palps Pair of appendages protruding from either side of mouthparts.

pathogen Anything that can produce disease.

pectoral fins Fins on either side of head.

pedipalp Second pair of appendages, between jaws and first pair of legs.

pelagic Occurring in open ocean.

pheromone Chemical released to attract member of opposite sex.

plankton Collective name of floating or drifting plants and animals, occurring in large masses of water.

poisonous Refers to substance that is harmful once ingested, inhaled or absorbed through skin.

posterior Rear of body.

proboscis Long and mobile feeding tube extending from front of head, seen in insects.

pronotum Plate-like structure covering all or part of dorsal surface of thorax of some insect species.

pupate Develop into a pupa.

species Basic unit of taxonomic classification.

spermatophore Sperm capsule.

spinneret Organ of a spider used to spin silk, located under abdomen.

spiracle Respiratory opening in exoskeleton of insects that allows air to enter.

spur Protruding growth of bone.

subspecies Level of taxonomic division below species.

substrate Underlying layer or material on which an organism lives.

symbiotic Refers to close, mutually beneficial relationship between two different organisms.

sympatry Refers to individuals that share same habitat.

taxonomy Classification of living things based on characteristics.

terrestrial Living on or spending time on the ground.

territory Area that an individual or group occupies and protects.

test Hard body covering of echinoderms.

torpor Period of physical inactivity (shorter than hibernation) induced by reduced body temperature and slowed metabolic rate.

toxin Poisonous substance produced within an animal or plant.

tubercle Rounded, raised nodule.

vector Carrier of disease.

venom Toxin secreted (and often injected) by an animal.

venomous Capable of secreting venom.

ventral Relating to undersurface or belly of an animal.

vertebral Along line of spine (vertebrae).

vertebrates Animals that have a backbone.

vestigial Forming a remnant of an appendage that has lost its original use through evolution.

viviparous Reproducing by giving birth to live young.

zoonotic Refers to type of infection or pathogen (zoonosis) that comes from an animal.

Australian Stinging Sponge ■ *Neofibularia mordens* W 40cm

DESCRIPTION Smooth sponge with raised ridges, being bright royal blue to purplish underwater, but becoming darker grey-brown with a blue wash when exposed to air.

Inside of sponge yellowish-brown.
DISTRIBUTION Recorded in southern Australian waters off SA and Tas.
HABITAT AND HABITS Found in coastal marine waters on rocky reefs, where it filters water through pores on the outer surface to extract nutrients. Produces a slime that causes contact dermatitis, which can lead to intense itchiness, swelling, a red or purple rash and severe pain, with some symptoms persisting for several weeks; if it occurs in the eye, it can lead to blindness. Experiments conducted in the early 1970s concluded that the dry sponge is not toxic. Harmful

Zoanthid ■ *Palythoa heliodiscus* DW 25mm

DESCRIPTION Broad, flat oral disc, usually pale to dark brown, with thin white lines radiating from oral opening, and with numerous small tentacles around outer rim.
DISTRIBUTION Recorded in tropical marine waters of Australia, primarily along eastern Qld. **HABITAT AND HABITS** Found in shallow waters, mainly at 4–24m depth, where it normally grows in dense colonies on coral and rocky reef substrates, with new polyps emerging from coenenchyme (mat) surrounding base of body (scapus). While there is no definitive proof that this species has the same toxicity as the closely related species *P. toxica*, studies have identified that it is the most similar in composition to the toxic zoanthids that have been available for testing. Poisonous

Feeding

At rest

Striped Anemone
■ *Dofleinia armata* DW 10cm

DESCRIPTION Large anemone, with broad basal disc, smooth column, and long cream to dark brown tentacles, some with darker longitudinal stripe, and slightly swollen tip. Tentacles and oral disc have numerous raised bumps, called papillae. **DISTRIBUTION** Tropical waters of Australia, from Perth WA, north through NT, to GBR Qld. **HABITAT AND HABITS** Found in sheltered areas around reefs and within mangroves, down to around 20m depth. Papillae on tentacles each contain nematocysts, which have venom that is used to immobilize prey or deter predators. Venom capable of inflicting an extremely painful injury to humans that can take many months to fully heal. The Striped Anemone has the ability to shed its tentacles as a defensive mechanism, which can make the surrounding waters problematic to divers. Venomous

Hell's fire anemone ■ *Actinodendron* sp. DW 20cm

DESCRIPTION More than one species is known as a hell's fire anemone, including *Actinodendron arboreum* and *A. plumosum*. Generally greyish and lacking any bright colouration, with numerous complex, branching tentacles, resembling broccoli in older individuals. Central oral disc has white stripes radiating from mouth. **DISTRIBUTION** Marine waters throughout Indo-Pacific. **HABITAT AND HABITS** Found in sheltered, shallow marine water down to 28m depth, on sandy and muddy substrates. The nematocysts can give an extremely powerful, intensely painful sting and cause skin ulceration in humans. More severe reactions can occur, including renal failure, anaphylactic shock and death. Venomous

Griffith's Fire Anemone ■ *Megalactis griffithsi* DW 20cm

DESCRIPTION Pale brown column and green to brown tentacles, with up to 35 relatively short secondary branches. Oral disc has complex pattern of green to brown and white,

with radiating lines. **DISTRIBUTION** Marine waters throughout Indo-Pacific. **HABITAT AND HABITS** Found in sheltered, shallow marine water down to 20m depth, on sandy and muddy substrates. Breeds by releasing eggs and sperm into the water. Once born the planktonic larvae are free swimming until they metamorphoze into juvenile sea anemones, at which stage they drop to the ocean floor and attach themselves to the substrate. The nematocysts can give an extremely powerful, intensely painful sting and cause skin ulceration in humans. More severe reactions can occur, including renal failure, anaphylactic shock and death. Venomous

Night Anemone ■ *Phyllodiscus semoni* Size highly variable

DESCRIPTION Variable, and capable of mimicry of corals. Typically a broad, translucent white basal disc, with orange centre, numerous branch-like pseudo-tentacles near base,

and ring of up to 160 long, translucent tentacles, each with faint white spots. **DISTRIBUTION** Marine waters of Central Indo-West Pacific. **HABITAT AND HABITS** Found on rocky or sandy substrates in association with reefs. During the day, spreads out pseudo-tentacles to obtain food from microscopic photosynthetic algae, but at night feeds on small invertebrates and plankton. The nematocysts of this anemone are capable of inflicting very painful injuries to humans, which are accompanied by swelling and ulceration of the affected area, and can take several months to heal. The toxin can also cause kidney damage, which can lead to death due to renal failure. Venomous

Box jellyfish ■ *Chironex fleckeri* BH 32.5cm
(Sea Wasp)

DESCRIPTION Box-shaped, transparent bell, with bunches of bluish-grey tentacles descending from corners (pedalia). **DISTRIBUTION** Tropical waters of northern Australia, from Exmouth WA, to Gladstone Qld. More widely spread in Indian and Pacific Oceans. **HABITAT AND HABITS** Young sessile polyps inhabit mangroves, and medusa stage moves into shallow creek systems and coastal waters before moving to inshore waters. Possibly restricted to waters less than 5m in depth, but a recent study has found members of this genus, perhaps including this species, over rocky reefs at depths of 56m. Feeds mainly on prawns, but also on fish. It has been responsible for 77 reported human deaths in Australia, and is most numerous during the warmer months of November–May. DANGEROUSLY VENOMOUS

Irukandji ■ *Carukia barnesi* BH 35mm

DESCRIPTION Translucent, with a small, elongated, box-shaped bell that is 2–3 times as tall as it is wide, and has a single tentacle on each lower corner. The nematocysts are found on the bell and tentacles. **DISTRIBUTION** Tropical waters of northern Australia, from around Broome WA, to Rockhampton Qld. Southern distributional limit is increasing due to rising water temperatures. **HABITAT AND HABITS** Found in deeper marine waters around reefs, where it occupies surface waters down to around 20m depth, and feeds on small fish and marine invertebrates. A sting from this species can be fatal to humans, producing Irukandji Syndrome, which manifests as limb and lower back pain, muscle cramping, headache, sweating nausea, vomiting, feeling of 'impending doom' and cardiac dysfunction. DANGEROUSLY VENOMOUS

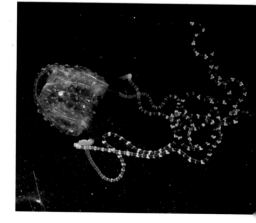

Jimble ■ *Carybdea rastonii* BH 60mm

DESCRIPTION As in other cubozoans, the bell of this species is cube shaped. It has four tentacles, one at each corner, and is translucent white, although the tentacles can

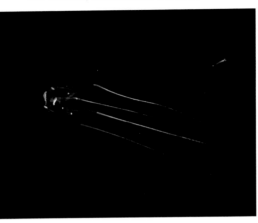

have a pinkish or purple tinge. **DISTRIBUTION** Waters of southern Australia, from Albany WA to southern Qld, and more widely in Pacific Ocean. **HABITAT AND HABITS** Found in colder marine and estuarine waters, where it occurs at depths from shallow surface waters to around 30m. During the day most common close to the substrate, but moves to surface waters between dusk and dawn. Preys on juvenile fish and zooplankton. Unlike more tropical species of cubozoan, the Jimble is not regarded as deadly, though its sting can cause intense localized pain and a red mark. Venomous

Box jellyfish ■ *Chiropsella bart* BH 50mm

DESCRIPTION Small, translucent, cube-shaped white bell with up to five long, thin (up

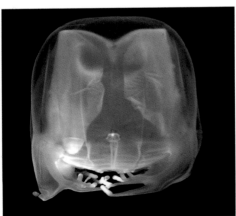

to 1.5mm diameter) tentacles on each pedalium (fleshy lobe on each of four lower corners of bell). Bell is smooth and lacks nematocysts. **DISTRIBUTION** Tropical waters of Gove Peninsula NT. **HABITAT AND HABITS** Found in shallow marine waters around sandy beaches, where it feeds on small crustaceans and fish, which are immobilized using venom contained in nematocysts on its tentacles. Occurs in larger numbers in August–December than at other times. Venom causes only mild localized pain and itchiness in most people. Venomous

Box jellyfish
■ *Chiropsella bronzie* BH 10.5cm

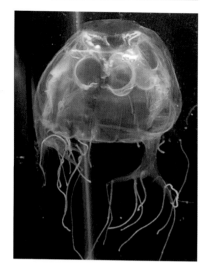

DESCRIPTION Small, translucent, cube-shaped white bell with up to nine long, thin (1mm diameter) tentacles on each pedalium (fleshy lobe on each of four lower corners of bell). Bell is smooth and lacks nematocysts. **DISTRIBUTION** Waters off north-eastern Qld, from Townsville to Cooktown. **HABITAT AND HABITS** Found in coastal marine waters around sandy beaches, where it feeds on small fish and crustaceans. Occurs in larger numbers during summer months than at other times, generally 14 days after significant rainfall. Venom causes only mild localized pain and itchiness in most people. Venomous

White-stinging Sea Fern ■ *Macrorhynchia philippina* H 45cm

DESCRIPTION Simple or clumping sea fern, with white to pale grey, feathery branches and dark brown to black stems. **DISTRIBUTION** In Australia, has been recorded from south-western WA, through

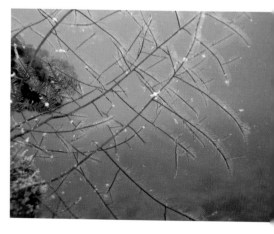

NT, to southern Qld. Occurs in warm temperate to tropical seas worldwide. **HABITAT AND HABITS** Found in shallow waters on coral and rocky reefs at depths of 3–25m, where it feeds on plankton. While it appears as a single organism, it is actually a colony of polyps called hydranths, each having a mouth (hypostome), tentacles, protective cup (hydrotheca) and stem (hydrocaulus), with the entire colony attached to the rocky or coral substrate by the hydrorhiza (rootstock) at the base. The hydranths can deliver a painful sting. Venomous

Fire coral ■ *Millepora platyphylla* H 50cm
(Hydrocoral)

DESCRIPTION Colours can range from cream to dark yellow or brownish, with paler edges to the large, plate-like colonies. Outer surface is calcified, with numerous small holes

(gastropores), which house softer digestive polyps (gastrozooids). **DISTRIBUTION** In northern Australia found from Houtman Abrolhos Island Group WA to GBR Qld. **HABITAT AND HABITS** Occurs in colonies on shallow marine coral reefs and adjacent slopes to around 18m depth. Relies on water currents to bring it prey, mainly plankton and small invertebrates, close enough for it to be immobilized with the nematocysts. Sting, although painful, is not considered dangerous to humans, usually producing only small welts, swelling and a rash. Some people may also suffer from nausea and vomiting, and potential allergic reactions or anaphylaxis. Venomous

Portuguese Man-of-War ■ *Physalia physalis* BW 150mm
(Bluebottle)

DESCRIPTION Generally blue, with enlarged translucent float, or pneumatophore, and numerous long, bead-like tentacles. **DISTRIBUTION** Tropical and subtropical Australia.

HABITAT AND HABITS Found in marine waters, where it moves along the surface of the water using its pneumatophore as a sail, which it can change in size and shape by regulating the amount of gas within it and, in doing so, influence its direction and speed of travel. Although it appears to be a single organism, this species is actually a colony of different kinds of polyp (zooids), which work together for survival. Tentacles are covered in nematocysts, which capture and immobilize prey, normally small fish, crustaceans and plankton. Can cause severe pain, skin sores, swelling, dizziness, vomiting and respiratory distress, and has been responsible for human fatalities. DANGEROUSLY VENOMOUS

Indo-Pacific Portuguese Man-of-War ■ *Physalia utriculus* BW 100mm
(Bluebottle)

DESCRIPTION Generally blue, with an enlarged translucent pneumatophore that can have a greenish or pinkish tinge, and long, dark blue main tentacle. **DISTRIBUTION** Around Australia. **HABITAT AND HABITS** Found in marine waters and often washed up on beaches. Moves along the surface of the water using its pneumatophore as a sail. Although it appears to be a single organism, this species is actually a colony of different kinds of polyps (zooids) that work together for survival. Captures and immobilizes prey, normally larval fish and small crustaceans, using nematocysts on its tentacles. Can cause intense localized pain and swelling in humans, and the potential for allergic reactions. Some published sources have synonimized this species with *P. physalis*. Venomous

Sea Nettle
■ *Chrysaora kynthia* BW 120mm

DESCRIPTION Somewhat flattened, translucent bluish-white bell, covered in colourless warts. Twenty-four long, flattened tentacles arranged in eight groups of three trailing from rim on underside. Long oral arms are thin and frilled, and hang below bell. **DISTRIBUTION** Small area of Indian Ocean between Rockingham and Perth WA. **HABITAT AND HABITS** Found in marine waters and estuaries, where it swims with the current as it hunts presumably for small fish, crustaceans, plankton and free-swimming larvae. The numerous nematocysts on the tentacles can deliver an intensely painful sting, accompanied by raised welts and redness, which may last for up to three days and can cause potential anaphylaxis in sensitive people. Venomous

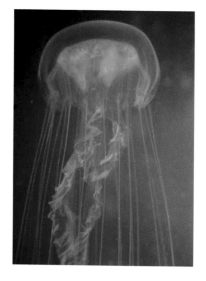

Mauve Stinger ■ *Pelagia noctiluca* BW 60mm

DESCRIPTION Variable in colour, including brownish-yellow, pale reddish or purplish, which is generally more intense in nematocysts and gonads, and luminesces when

disturbed. Mushroom-shaped bell has 16 outer lobes, with eight thin marginal tentacles. Undersurface has four thick oral arms. **DISTRIBUTION** Marine waters worldwide and throughout Australia. **HABITAT AND HABITS** Found in warm and temperate waters, where it usually forms large swarms that feed on other small jellyfish, and free-swimming larval sea squirts and tunicates. Stinging cells are capable of causing intense and immediate pain when contacted by the skin, accompanied by tissue damage, swelling and blood cell damage. Symptoms are generally short lived. Venomous

Lion's Mane Jellyfish ■ *Cyanea capillata* BW 80cm

DESCRIPTION Translucent, disc-shaped bell divided into eight distinct lobes, with eight groups of up to 150 long, fine trailing tentacles descending from underside. Large oral

arms on undersurface have red or yellow tentacles. **DISTRIBUTION** In Australia, members of this genus have been recorded along all the coast, although the exact distribution of the species is unclear. Considered to be mainly from southern Qld to southern NSW. **HABITAT AND HABITS** Occurs in coastal and deeper marine waters, where its long stinging tentacles ensnare fish (mainly), crustaceans and other marine animals. Stinging cells on tentacles and medusa can be very painful if touched, and accompanied by long-lasting blisters and irritation. Sensitive people can exhibit muscle cramping, heart problems and respiratory distress. Venomous

Southern Blue-lined Octopus ■ *Hapalochlaena fasciata* TL 15cm

DESCRIPTION Small with short arms, and generally cream to orange with numerous blue lines on mantle, and blue rings on arms and arm crown. When agitated, blue lines and rings become more visible.

DISTRIBUTION Subtropical waters of eastern Australia, from southern Qld to southern NSW. **HABITAT AND HABITS** Found in shallow intertidal waters near rocky reefs to around 20m depth. Mainly nocturnal. Prey consists mostly of small crustaceans, which are immobilized with a powerful neurotoxin contained within the organs and skin of the octopus, and injected using modified salivary glands. If handled, can inject venom by biting through the skin, or venom can be transferred to the bloodstream by the person licking their fingers. One human death has been attributed to this species. DANGEROUSLY VENOMOUS

Southern Blue-ringed Octopus ■ *Hapalochlaena maculosa* TL 22cm

DESCRIPTION Small and short armed, and generally greenish to cream with numerous darker blotches and bands, each containing small blue ring. When agitated or threatened,

rings become brighter and body becomes darker. **DISTRIBUTION** Waters of southern Australia, from eastern Vic, through Tas and SA, to south-western WA. **HABITAT AND HABITS** Found in rocky reef areas and seagrass meadows in shallow water to around 50m depth, and in rock pools. Hunts at night for crabs, which it paralyses by spraying a toxic saliva into the water. Venom is highly toxic to humans and, if picked up by an unsuspecting person, the octopus can bite them and pump the venom through the open wound. DANGEROUSLY VENOMOUS

Northern Australian Greater Blue-ringed Octopus
■ *Hapalochlaena lunulata* TL 22cm

DESCRIPTION Generally dark yellow or brown to grey, with numerous darker blotches and bands, each containing large blue ring. When agitated or threatened, rings become

brighter and body becomes darker. **DISTRIBUTION** Coastal waters of northern Australia, and north throughout warmer waters of Western Pacific Ocean. **HABITAT AND HABITS** Found in shallow rocky reef areas and small pools within intertidal zone, where it shelters during the day in crevices, large seashells or discarded items. Hunts along the sea bottom for crabs, which it paralyses by spraying a toxic saliva into the water. Envenomation from this species resulted in the first documented human fatality from this genus. DANGEROUSLY VENOMOUS

Poison Ocellate Octopus ■ *Amphioctopus mototi* TL 32cm
(Mototi Octopus; fe'e mototi [Rapa Iti Island])

DESCRIPTION Variable. Able to change colours and patterns depending on its mood, but normally cream and orange-brown, changing to white with purplish-red stripes when

threatened. Iridescent blue circle also appears on each side when alarmed. **DISTRIBUTION** Tropical waters of northern NSW and GBR Qld, and throughout South Pacific Ocean. **HABITAT AND HABITS** Inhabits sandy areas adjacent to rocky coral rubble at 1–54m depth. Feeds presumably at night on shellfish and hermit crabs, which are first paralysed using a powerful saliva, which it injects into the shell of its prey after drilling a small hole. Shelters in deep lair excavated beneath coral rubble or under rocky overhangs. Aggressive, and readily bites intruders. Venomous

Pfeffer's Flamboyant Cuttlefish

■ *Metasepia pfefferi* TL 80mm (mantle 60mm)

DESCRIPTION Colour normally dark brown to purple with variable yellow-and-white patterning, but changes colour rapidly when threatened, displaying warning colours of black, red (tips of arms) and yellow. **DISTRIBUTION** Tropical waters of northern Australia, from Mandurah WA, through NT, to Moreton Bay Qld. Also southern PNG. **HABITAT AND HABITS** Found in coastal and offshore waters at depths of 3–86m, where it actively hunts during the day for fish and crustaceans. Thought to be the only cuttlefish to 'walk' on the ocean floor. The tissues contain a potent toxin that can be deadly to humans if a sufficient quantity is consumed. Poisonous

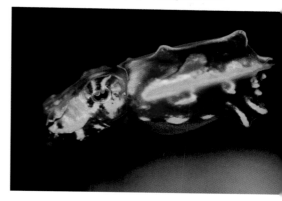

Striped Pyjama Squid ■ *Sepioloidea lineolata* TL 70mm

DESCRIPTION Variable. Normally white with conspicuous dark reddish-brown to purplish-black longitudinal stripes, but can have varying degrees of dark purplish-brown. Upper portion of eye covering is yellowish. **DISTRIBUTION** Widespread in marine waters around southern Australia, from around Brisbane Qld to Shark Bay WA. **HABITAT AND HABITS** Found in coastal waters to around 20m depth, where it conceals itself during the day beneath the sand with only its eyes visible. Any prey, mainly fish and small crustaceans, which comes within range is seized using a pair of extendable tentacles. Secretes copious amounts of toxic slime from glands on sides and ventral surface when threatened. It is thought that the striking body colouration is an indication that it is toxic. Poisonous

Court Cone ▪ *Conus aulicus* TL 163mm

DESCRIPTION Narrow, cylinder-shaped shell, with moderately large opening for single muscular foot and head to protrude. Chocolate-brown to reddish-brown, with strongly demarcated, variably sized, round to triangular white spots. **DISTRIBUTION** Northern Australia from north-western NT to south-eastern Qld. Widely distributed in tropical and subtropical marine waters of Indo-Pacific region. **HABITAT AND HABITS** Found

in coastal waters, on sandy substrates, generally to a depth of 30m. Feeds on other gastropods (mostly snails), which are immobilized with venom-soaked darts injected using a propelled radular tooth. The helpless victim is then ingested through the expanded proboscis. DANGEROUSLY VENOMOUS

Captain Cone ▪ *Conus capitaneus* TL 98mm

DESCRIPTION Broad, conical shell with wide white aperture. Variable in colour and pattern, although usually pale yellowish to orange-brown, with diagonal paler band,

spiralling dashed dark brown lines, and low white and dark brown spire. **DISTRIBUTION** Northern Australia from north-western WA, through NT and Qld, to north-eastern NSW. Widely distributed in tropical and subtropical Indo-Pacific region. **HABITAT AND HABITS** Occurs from intertidal zone to deeper marine waters to 240m depth, on wide range of substrates, including coral reef flats to sand, where it shelters among algae or in rocky crevices. Feeds on polychaete worms, which are immobilized with venom-soaked darts injected using a propelled radular tooth, and ingested through the expanded proboscis. Venomous

Cat Cone ▪ *Conus catus* TL 52mm

DESCRIPTION Broad, conical shell with wide opening for single muscular foot and head to protrude. Variable in colour and pattern, although usually creamish with brownish blotches. **DISTRIBUTION** Northern Australia from north-western WA, through NT, to south-eastern Qld. Widely distributed in tropical and subtropical marine waters of Indo-Pacific region. **HABITAT AND HABITS** Found in intertidal zone on range of marine substrates, from coral reefs to sand, generally to a depth of 8m, where it feeds on small fish. Once located, prey is immobilized with venom-soaked darts, injected using a propelled radular tooth. The helpless victim is then ingested through the expanded proboscis. DANGEROUSLY VENOMOUS

Geographer Cone ▪ *Conus geographus* TL 150mm

DESCRIPTION Hard, smooth shell, with numerous whorls and large opening about a third the width of main shell body. Base shell colour cream, white or pale pink, with variable amounts of reddish-brown markings forming longitudinal spiral pattern. **DISTRIBUTION** In Australia, from Carnarvon WA, through northern Australia, to around Brisbane Qld. Generally in tropical and subtropical waters of Indo-Pacific region. **HABITAT AND HABITS** Primarily inhabits sandy intertidal zones and coral reefs, where it hunts at night for fish, which are detected using chemical receptors housed in a modified section of the mantle tissues called a siphon. Once detected, fish are harpooned with paralysing, venom-soaked darts and swallowed whole using the extendable proboscis. The only confirmed death in Australia from a cone snail sting was from this species. DANGEROUSLY VENOMOUS

Leopard Cone ▪ *Conus leopardus* TL 200mm

DESCRIPTION Broad, conical shell, with narrow opening for single muscular foot and head to protrude, and flattened spire. Base colour off white, with spiralled pattern of small, dark brown to greyish blotches. **DISTRIBUTION** Northern Australia, from north-western WA, through NT, to south-eastern Qld. Widely distributed throughout tropical marine waters of Indo-Pacific region. **HABITAT AND HABITS** Found mostly on reef flats in shallow bays to 45m depth, usually on rocky or sandy substrates, where it feeds on worm-like marine invertebrates (hemichordates). Once located, prey is immobilized with venom-soaked darts, injected using an extendable, hollow proboscis, then ingested through the expanded proboscis. Venomous

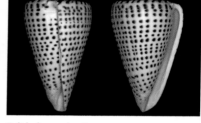

Unpolished cone in natural setting *Polished cone showing aperture*

Magical Cone ▪ *Conus magus* TL 94mm

DESCRIPTION Broad, cylinder-shaped shell with wide aperture. Variable in colour and pattern, although usually white with orange-brown to dark brown blotches, and with faint

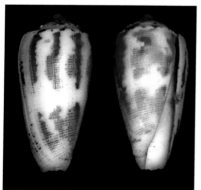

regular pattern of dark brown dotted lines. **DISTRIBUTION** In Australia, occurs off NT and Qld. Widely distributed in tropical and subtropical marine waters of Indo-Pacific region and Red Sea. **HABITAT AND HABITS** Found in shallow waters on range of marine substrates, from rocky coral reefs to sand, to around 100m depth, where adults feed on small fish and young on marine worms. Hunts nocturnally and once located, prey is immobilized with venom-soaked darts. Venom from this species has been used to produce a pharmaceutical painkiller called ziconotide, which is much more powerful than morphine. DANGEROUSLY VENOMOUS

Marbled Cone ■ *Conus marmoreus* TL 150mm

DESCRIPTION Broad, conical shell with moderately large aperture, and reticulated pattern of orange to black with small spots, to large, subtriangular blotches of white. **DISTRIBUTION** Northern Australia from north-western WA, through NT, to south-eastern Qld. Widely distributed in tropical and subtropical marine waters of Indo-Pacific region. **HABITAT AND HABITS** Found in shallow marine waters on rocky coral reef flats and adjacent sandy patches, generally to 30m depth, where it shelters under rocky ridges, within weed or under sand. Generally active during rising tide, but some individuals hunt throughout the day. Preys on other gastropods, which are immobilized with venom-soaked darts, injected using a propelled radular tooth, then ingested through the expanded proboscis. DANGEROUSLY VENOMOUS

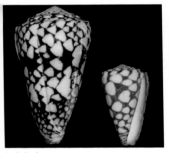

Polished cones

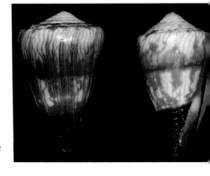

Cone with proboscis extended

Soldier Cone ■ *Conus miles* TL 132mm

DESCRIPTION Broad, conical, thickened shell, with shallow spire and moderately wide, brown-and-white aperture. Variable colouration, although base colour is generally white to yellowish-orange, with dark brown diagonal band, dark brown base and numerous reddish-brown fine, longitudinal wavy lines. **DISTRIBUTION** Northern Australia from central coast of WA, through NT and Qld, to north-eastern NSW. Widely distributed in tropical and subtropical marine waters of Indo-Pacific region. **HABITAT AND HABITS** Occupies intertidal and subtidal zones in shallow marine waters to around 50m depth, on either rocky or sandy substrates, where it feeds on marine worms. Prey is immobilized with venom-soaked darts, injected using a propelled radular tooth, then ingested through the expanded proboscis. Venomous

Pearled Cone ▪ *Conus omaria* TL 85mm

DESCRIPTION Long, conical shell, with moderate spire and wide white aperture. Typically white with network of orange-brown to chocolate-brown lines forming range

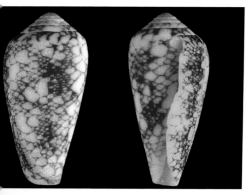

of white spots and larger subtriangular blotches. **DISTRIBUTION** Northern Australia from north-western WA, through NT, to south-eastern Qld. Widely distributed in tropical and subtropical marine waters of Indo-Pacific region. **HABITAT AND HABITS** Found on rocky reef flats in marine waters of subtidal zone at 10–100m depth, where it hunts mainly at night in algal patches for other gastropods. Once located, these are immobilized with venom-soaked darts and ingested through the expanded proboscis. During the day shelters in crevices, under rocks or buried in sand. Venomous

Striated Cone ▪ *Conus striatus* TL 129

DESCRIPTION Broad, cylinder-shaped shell, with wide opening for single muscular foot and head to protrude. Generally pinkish-white with brownish blotches and lines. **DISTRIBUTION** In Australia, occurs from north-western WA, through NT, to south-eastern Qld. Widely distributed in tropical and subtropical marine waters of Indo-Pacific region and Red Sea. **HABITAT AND HABITS** Found in shallow marine waters, usually on coral reef and sandy substrates, generally to 25m depth, where it feeds on small fish. Once located, prey is immobilized with venom-soaked darts, injected using a propelled radular tooth. The helpless victim is then ingested through the expanded proboscis. DANGEROUSLY VENOMOUS

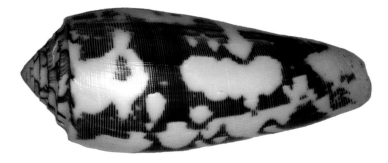

Textile Cone ■ *Conus textile* TL 150mm
(Cloth-of-gold Cone)

DESCRIPTION Glossed, broad, conical shell with wide opening. Base colour a mix of yellowish-brown and brown, with variable pattern of dark-edged, subtriangular, 'shingle-like' white markings and dark brown lines.

DISTRIBUTION Northern Australia from north-western WA, through NT and Qld, to north-eastern NSW. Widely distributed in tropical and subtropical marine waters of Indo-Pacific region. **HABITAT AND HABITS** Found in shallow marine waters, in open sandy, muddy and coral substrates, generally to 50m depth. Feeds mainly on other gastropods, small fish and marine worms, which are immobilized with venom-soaked darts, then ingested through the expanded proboscis. Human deaths have been recorded for this species, although none in Australia. DANGEROUSLY VENOMOUS

Tulip Cone ■ *Conus tulipa* TL 95mm

DESCRIPTION Broad, cylindrical shell with very wide opening that is bright violet in some individuals. Generally purplish-white with reddish-brown to chestnut blotches and dashed lines. **DISTRIBUTION** Northern Australia from around Christmas Island WA, through NT, to south-eastern Qld. Widely distributed in tropical and subtropical marine waters of Indo-Pacific region. **HABITAT AND HABITS** Found in shallow waters on a range of marine substrates, including weedy coral flats and rocky reefs, down to 10m depth, where it feeds on small fish. Prey is immobilized with venom-soaked darts, then swallowed whole using the expanded proboscis. DANGEROUSLY VENOMOUS

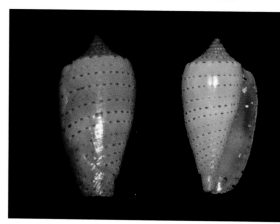

Vexillium Cone ▪ *Conus vexillum* TL 183mm
(Flag Cone)

DESCRIPTION Broad, conical shell with moderately shallow spire and wide white aperture. Variable in colour and pattern, although usually yellowish to reddish-brown with

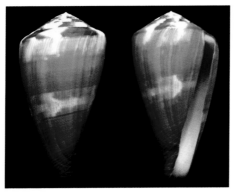

obscure diagonal white central band and striated, white-blotched spire. **DISTRIBUTION** Northern Australia from north-western WA, through NT and Qld, to north-eastern NSW. Widely distributed in Indo-Pacific region. **HABITAT AND HABITS** Found in intertidal (as juveniles) and subtidal (as adults) zones of shallow marine waters down to 70m depth, where it hunts on rocky or sandy substrates for eunicid marine worms. Prey is immobilized with venom-soaked darts, then ingested through the expanded proboscis. Venomous

Sea Swallow ▪ *Glaucus atlanticus* TL 3cm

DESCRIPTION Usually silvery-grey above and bright blue below, with dark blue longitudinal stripes. Six limbs, each with numerous long, finger-like appendages (cerata)

along body. **DISTRIBUTION** Indian, Pacific and Atlantic Oceans. **HABITAT AND HABITS** Found in temperate and tropical waters, where it is often seen in close association with members of the cnidarians, including the Portuguese man o'wars, on which it feeds. Recycles its victim's nematocysts and stores them in cnidosacs on the ends of its cerata for its own self-defence. People can have mixed reactions from stings, depending on the concentration of nematocysts it is carrying, with pain, swelling and redness varying in intensity and lasting from a few minutes to several hours or more. Venomous

Blue Sea Slug ■ *Glaucus marginatus* TL 1cm
(Margined Sea Lizard)

DESCRIPTION Silvery-grey above and dark blue below, with darker blue stripe along ventral surface, edged with glittery blue flecks. Six limbs, each with single row of cerata in fan-shaped arch at extremities of limbs.

DISTRIBUTION Pacific and Atlantic Oceans. **HABITAT AND HABITS** Found in pelagic waters, where it is often seen floating upside-down on the surface. Feeds on cnidarians, including the Portuguese man o'wars, recycling its victim's nematocysts and storing them in cnidosacs on the ends of its cerata for its own self-defence. People can have mixed reactions from stings, depending on the concentration of nematocysts it is carrying, with pain, swelling and redness varying in intensity and lasting from a few minutes to several hours or more. Venomous

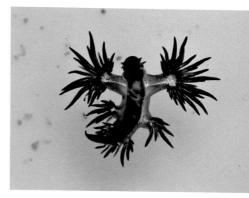

Orange Fireworm ■ *Eurythoe* cf. *complanata* TL 35cm
(Common Fireworm)

DESCRIPTION Long, flattened, segmented body. Each segment, with the exception of the head and last segment, has a pair of fleshy appendages and brittle white bristles. Colour variable, being orange, pinkish or bluish-green. **DISTRIBUTION** In Australia, individuals referred to as this species occur from south-western WA, through the tropical north, to around Jervis Bay NSW. **HABITAT AND HABITS** Found in warm, shallow waters of intertidal zone, on rocky reefs and adjacent sandy or muddy areas, where it shelters under rocks, emerging mainly at night to forage for carrion. The white bristles can easily pierce human skin (and dive gloves) and break off, with the venom at the base producing intense localized burning pain. Three or more distinct species are potentially currently included in this species complex. Venomous

Bearded Fireworm ■ *Hermodice* cf. *carunculata* TL 10cm

DESCRIPTION Long, flattened, segmented body. Each segment, with the exception of the head and last segment, has a pair of fleshy appendages, red gills and brittle white bristles. Other body colours include

yellow and green. **DISTRIBUTION** Widespread in Western Central Atlantic, Mediterranean Sea and Indo-Pacific. Almost certainly a suite of cryptic species, originally described from the West Indies. **HABITAT AND HABITS** Found along coral and rocky reefs, and adjacent sandy or muddy areas, where it shelters under rocks, emerging mainly at night to forage for corals, anemones and small crustaceans. Bristles are hollow and contain a venom at the base that produces an intense localized burning pain. Venomous

Australian Land Leech ■ *Chtonobdella limbata* TL 80mm

DESCRIPTION Brownish to blackish, with paler longitudinal stripe along anteriorly tapering body. It has two jaws and leaves a V-shaped mark after biting. **DISTRIBUTION**

Restricted to coast and near inland of southern NSW, although isolated records have recently come from the border of northern NSW and southern Qld. **HABITAT AND HABITS** Found in forested areas, where it occurs on or close to the ground, and is generally most active following rain. Feeds on blood of mammals, including humans, using its jaws to pierce the skin of its host and injecting a secretion called hirudin to prevent the blood from clotting. The wound can bleed and stay itchy and inflamed for several hours after the leech has dropped off. Harmful

Australian Paralysis Tick ■ *Ixodes holocyclus* TL 4mm (before feeding)
(Grass Tick; Bush Tick; Seed Tick)

DESCRIPTION Generally flattened, with eight legs (adults), and can grow to the size of a pea after feeding. Colour variable, from yellowish-brown to greyish. **DISTRIBUTION** Eastern Australia, east of Great Dividing Range. **HABITAT AND HABITS** Found in moist habitats, including wet sclerophyll forests, temperate rainforests, woodland, grassland, and adjacent parks and gardens, where it is most active at times of high humidity or following rains. Most tick bites cause only minor swelling, itchiness and redness, but some people can develop more severe reactions, including anaphylaxis, breathing difficulty, facial paralysis, sensitivity to bright light, dizziness and fever. Potentially Dangerous

Before a blood meal

Fully engorged with eggs

Salt Lake Scorpion ■ *Australobuthus xerolimniorum* TL 45 mm

DESCRIPTION Pale yellowish-brown, with small pedipalps and small aculeus (stinger). Eyes are black. Female slightly larger than male and more thickset. **DISTRIBUTION** Around salt lake areas of SA. **HABITAT AND HABITS** Occurs along margins of salt lakes, where it shelters during the day under vegetation litter or in soil cracks, and emerges on warm nights to forage on hard lake surface for spiders and flies. Stings are generally not considered lethal to humans, producing intense pain, sweating, headaches and nausea. Venomous

Spider-hunting Scorpion ■ *Isometroides vescus* TL 55mm
(Spiral Burrow Scorpion)

DESCRIPTION Pale yellowish-brown to greyish-brown, with small pedipalps and dark brown tail-tip and aculeus (stinger). Legs are yellowish, occasionally with indistinct spotting. **DISTRIBUTION** Widespread mainly in inland areas of southern Australia, but also along southern and western coast, from central western WA, to north-western Vic

and western NSW. **HABITAT AND HABITS** Found in drier areas, where it shelters during the day in burrows of burrowing spiders, which it specializes in hunting, and emerges on warmer nights to hunt. Venom causes localized pain in humans, accompanied by swelling. Venomous

Buchar's Scorpion
■ *Lychas buchari* TL 50mm

DESCRIPTION Pale yellow to yellowish-brown and greyish, with small pedipalps and greyish tail-tip and aculeus (stinger). Legs are yellowish to greyish. **DISTRIBUTION** Central and north SA, and western WA. **HABITAT AND HABITS** Found in mallee on sandy dunes around salt lakes. Hunts on the surface at night for insects and other invertebrates. During the day shelters under loose bark. Venom causes intense pain, sweating, headaches and nausea. Venomous

Marbled Scorpion ■ *Lychas marmoreus* TL 40mm

DESCRIPTION Small scorpion. Generally yellowish-brown to dark greyish, with darker marbled pattern. Pedipalps are small. **DISTRIBUTION** Throughout southern Australian mainland, from south-western WA to south-eastern Qld. **HABITAT AND HABITS** Found in eucalypt forests and adjacent areas, where it prefers cooler moist areas, and occasionally enters houses. Hunts at night for small insects, and shelters during the day under loose bark and rocks, or in leaf litter. This species is responsible for most scorpion stings on humans. The venom can cause severe pain and inflammation that can persist for several hours, and allergic reactions in sensitive people. A single human fatality, a small infant, has been attributed to this species. Venomous

Forest Scorpion ■ *Cercophonius squama* TL 35mm

DESCRIPTION Variable, from yellowish-brown to dark brown with reddish-brown patterning. Body moderately thickset, with small pedipalps. **DISTRIBUTION** Widespread in southern eastern Australia, from south-eastern SA to eastern NSW, and an isolated population in south-western WA. The only scorpion species in Tas. **HABITAT AND HABITS** Occurs in moist habitats, where it lives in a burrow, emerging at night to feed on small invertebrates. Female gives birth to live young, which are carried on her back for the first few weeks of life. Capable of inflicting a painful sting, which is accompanied by swelling and redness that can last some hours. A single human fatality, a small infant, has been attributed to this species. Venomous

Rainforest Scorpion ■ *Hormurus waigiensis* TL 80mm

DESCRIPTION Uniformly dark brown, with large, dark greyish pincers (pedipalps) and short, slender tail. **DISTRIBUTION** Coastal Qld and north-eastern NSW. **HABITAT AND HABITS** Found in forests and rainforests, where it shelters during the day in

crevices in rocks and logs, under bark or in burrows. Emerges at night to hunt for insects and other invertebrates, which are caught and often killed using the large pincers. Will also opportunistically seize prey if it comes close enough to its shelter. Female gives birth to around 20 live young, which are carried on her back for several weeks. Venom has only been reported to have mild effects on humans, but the large pincers can inflict a painful 'bite'. Venomous

Flinders Ranges Scorpion ■ *Urodacus elongatus* TL 120mm

DESCRIPTION One of the largest scorpion species in Australia, with males reaching a larger size than females, and having longer tail segments. Yellowish to brownish and grey, with blackish claws.

DISTRIBUTION Confined to Flinders Ranges region SA. **HABITAT AND HABITS** Inhabits gullies and similar lowland places in rocky areas. Shelters during the day in a burrow and under rocks, and hunts at night by sitting and waiting at entrance of burrow for prey to come within range. Males are wider roaming than females, and often do not have a resident burrow. Sting can be painful, and accompanied by swelling and redness. Venomous

Black Rock Scorpion ■ *Urodacus manicatus* TL 55mm

DESCRIPTION Yellowish-brown or reddish-brown to dark brown, with four pairs of legs; the front pair have become modified into a pair of large pincers, or pedipalps. **DISTRIBUTION** Eastern Australia, including Qld, NSW, Vic and SA. **HABITAT AND HABITS** Found in open forests and woodland, where it lives in a shallow burrow under a rock or log, about 10cm deep and with the entrance under a small rock. Ambush predator of beetles, cockroaches, myriapods, spiders and small reptiles, waiting near entrance of its burrow for prey to come within range. Prey subdued with venom injected using stinger on tip of tail. Sting can be painful, and accompanied by swelling and redness. Venomous

Under natural light

Under ultra-violet light

Desert Scorpion ■ *Urodacus yaschenkoi* TL 80–110mm

DESCRIPTION Variable within range, being yellow or reddish-brown, with moderately sized pedipalps and aculeus (stinger). **DISTRIBUTION** Widespread throughout interior of Australia, from western NSW to north-western WA. **HABITAT AND HABITS** Found in woodland and shrubland in arid sandy deserts, where it shelters during the day in deep, spiral-shaped burrows within an open area. Waits at entrance of burrow at night to ambush prey, usually insects and other invertebrates, which are grabbed and often killed with the large pincers, but can also be subdued with venom injected using the aculeus on the tail-tip. Sting can be painful, accompanied by mild swelling. Venomous

Black House Spider ■ *Badumna insignis* TL 19mm

DESCRIPTION Dark brownish or greyish to blackish, with white-flecked pattern on dorsal surface. Female has large, bulbous abdomen, while male's is thinner and tapers towards

tip. **DISTRIBUTION** Throughout subtropical Australia, from south-eastern Qld, to south-eastern Vic, Tas, Adelaide region SA and Perth region WA. **HABITAT AND HABITS** Found in dry forests and woodland, but also around windows and wall crevices in houses and other structures, on garden trees, and around rock walls and logs, where it builds large, intricate woolly webs on the surface, with a funnel retreat. Feeds on flies, moths, bees and beetles. Bites in humans, although uncommon, can result in serious symptoms, including severe pain, inflammation, sweating, dizziness, nausea and vomiting. Venomous

Slender Sac Spider
■ *Cheiracanthium gilva* TL 18mm

DESCRIPTION Generally pale yellowish to brownish. Slender body and long legs; front two pairs of legs long with black tips and often held forwards, and rear pair face rearwards. **DISTRIBUTION** Throughout Australia. **HABITAT AND HABITS** Inhabits vegetated areas, where it actively hunts in bushes and shrubs for insects, including butterflies. Builds small, cylindrical retreats. Its bite, although uncommon, can produce a range of symptoms and some longer term conditions have been reported, but normal symptoms include minor pain, swelling and redness. Venomous

Common White-Tailed Spider ■ *Lampona cylindrata* TL 20mm

DESCRIPTION Long, cylindrical, dark greyish body, with two pairs of greyish-white spots and large whitish tip on abdomen, and reddish-brown pointed legs. Females are larger than males. **DISTRIBUTION** Throughout southern Australia from south-eastern Qld, through NSW, Vic, Tas and SA, to WA. **HABITAT AND HABITS** Mainly occurs in vegetated areas, including on trees, and under logs, rocks and leaf litter. Hunts at night, mostly for other spiders, but will readily enter houses. Bites can be painful and accompanied by swelling, redness and itchiness lasting up to 12 days. They can become infected with the bacterium *Mycobacterium ulcerans*, which causes tissue necrosis but is incorrectly believed by many people to be caused by the venom itself. Venomous

White-tailed Spider ■ *Lampona murina* TL 20mm

DESCRIPTION Long, cylindrical, dark greyish body, with two pairs of whitish spots and large whitish tip on abdomen, and reddish-brown pointed legs. Females are larger than males.

DISTRIBUTION Eastern Australia, from north-eastern Qld, through eastern NSW and north-eastern Vic. **HABITAT AND HABITS** Occurs in urban areas, around houses and in parks and gardens, where it can be found on trees, under logs, rocks and leaf litter, and among human refuse. Hunts at night, mainly for other spiders, including the Redback Spider (see p. 53). Some bites result in necrotizing arachnidism, caused by infection of the bite site with the bacterium *Mycobacterium ulcerans*. Venomous

Garden Wolf Spider
■ *Tasmanicosa godeffroyi* TL 35mm

DESCRIPTION Predominantly brown, with paler yellowish-brown to orange-brown and darker blackish lines and marks, and with characteristic radiating pattern on carapace. Females are larger than males. **DISTRIBUTION** Widespread throughout temperate Australia. **HABITAT AND HABITS** Occurs in open areas in a variety of wooded and grassland habitats, including gardens. Lives in a lidless burrow, the entrance of which is lined with silk, and has a small, raised rim of leaves, sticks and grasses. Nocturnal, terrestrial hunter, actively pursuing and preying on small to large insects and other spiders. Although non-aggressive, it is still capable of giving a painful bite that can become easily infected. Venomous

Green Jumping Spider ■ *Mopsus mormon* TL 16mm

DESCRIPTION Yellowish and green, with reddish-brown and white patch on cephalothorax. Two large black eyes and several widely spaced smaller eyes, and large

fangs. Males have a blackish face, bordered by greyish-white whiskers. Legs have reddish-brown blotches and abdomen has two longitudinal blackish lines. **DISTRIBUTION** Northern Australia, from Kimberley region WA, through northern NT and eastern Qld, to around Coffs Harbour NSW. **HABITAT AND HABITS** Found in vegetated areas with green leaves, where it actively hunts on foliage during the day for insects and other spiders. Prey is seized by an accurate leap from several centimetres away, often with a silken lifeline attached to prevent the spider from falling to the ground in the event of it losing its footing. Defensive spider that will readily bite humans. Bites can be severely painful. Venomous

Beautiful Badge Huntsman ■ *Neosparassus calligaster* TL 20mm

DESCRIPTION Large, long-legged spider with conspicuous shield-shaped 'badge' on underside of abdomen. Generally orange-brown to dark brown above, with darker flecking, and two larger blackish spots on dorsal surface of abdomen. Legs have alternate blackish and either yellowish or white bands, and areas of yellow, with fine black speckling. **DISTRIBUTION** Coastal and inland eastern Australia, from around Mackay Qld, through NSW, to Vic and Tas. Isolated records from WA (possibly vagrants). **HABITAT AND HABITS** Occurs in range of habitats, where it shelters under loose bark and leaf litter, in crevices and under rocks, emerging to hunt at night for insects and other invertebrates. Bites attributed to this species have occurred, with symptoms including severe pain, inflammation, sweating, nausea and vomiting. Venomous

Ventral surface showing 'badge' *Dorsal pattern*

Redback Spider ■ *Latrodectus hasselti* TL 10mm

DESCRIPTION Female usually black, but can also be brownish, with an orange-red blaze on the dorsal surface of the abdomen, and a similarly coloured hourglass shape on the undersurface. Male much smaller (3–4mm), pale brownish, with white spots and streaks, and with an hourglass pattern on the undersurface of the abdomen. **DISTRIBUTION** Australia-wide in close proximity to human habitation. **HABITAT AND HABITS** Found in relatively dry terrestrial areas, often in disused building materials, around houses and in gardens. Constructs tangled web of vertical threads that have a sticky coating designed to entangle prey – usually insects, other spiders and small lizards. This is then hoisted up into the air before being bitten and consumed, or stored for later consumption. Due to this spider's close association with human habitation its bites are common, although significant envenoming only occurs in about 20 per cent of bites. Symptoms can include pain, sweating, nausea, vomiting, muscle weakness and possibly death (particularly in the aged and very young), without the administration of antivenom. Redback Spider Antivenom was introduced in 1956. DANGEROUSLY VENOMOUS

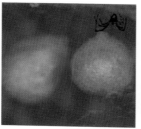

Much smaller male and egg sacs *Female*

Eastern Mouse Spider ▪ *Missulena bradleyi* TL 25mm

DESCRIPTION Stocky, uniformly black spider, with enlarged cephalothorax and basal part of fangs. Male has whitish or pale blue spot on abdomen. Male smaller than female, with

comparatively longer legs and only slightly swollen palps. **DISTRIBUTION** Along coasts and eastern slopes of eastern Australia. **HABITAT AND HABITS** Occurs in forests and woodland, where female occupies a hole in the ground (with a trapdoor lid), while male is most often encountered roaming in search of a mate. Prey animals include insects, other spiders, small lizards and frogs. Although the majority of bites do not result in significant envenomation and generally only cause minor to moderate effects, including localized pain, swelling, burning and itchiness, severe envenomation has been recorded in a young child. Venomous

Red-headed Mouse Spider ▪ *Missulena occatoria* TL 15–35mm

DESCRIPTION Male more colourful than female, with red on enlarged cephalothorax and basal part of broad fangs, and dark metallic blue on abdomen. Female larger than male, dark brown to black (occasionally with reddish wash on jaws), and with shorter legs. **DISTRIBUTION** Throughout mainland Australia, west of Great Dividing Range. **HABITAT AND HABITS** Found in open forests, woodland and dry shrubland, where it hunts at night for insects, other spiders, small lizards and frogs. Female lives in a burrow (with a lid) in an earthen bank near a waterway, while male is most often encountered roaming in search of a mate. Most bites cause only localized inflammation, burning and itchiness. Venomous

Female

Brightly marked male

Common Brown Trapdoor ■ *Misgolas gracilis* CL 8.82mm

DESCRIPTION Generally dark brown, usually with paler brown bands on abdomen, and with fine covering of paler brown hairs on carapace. Spinnerets are short and conical. Male more slender than female, with longer legs and enlarged 'boxing glove' shaped palps. **DISTRIBUTION** Wide ranging throughout NSW, in region bounded by Tamworth (north), Mudgee (south), Jamberoo (east) and Warrumbungle Range (west). **HABITAT AND HABITS** Found in variety of habitats. Female lives in a silk-lined burrows in an earthen bank along a stream and in areas of open ground; burrow is concealed by a thin, fragile lid across the opening. Males are most usually encountered as they wander around suburban parks and gardens during the mating season. Venomous

Male has 'boxing glove' palps

Female is larger with shorter legs

Sydney Brown Trapdoor ■ *Misgolas villosus* CL c. 10mm

DESCRIPTION Generally dull and hairy, yellowish-brown to dark brown on carapace, and brown with obscure grey bars on abdomen. Male has distinctive 'boxing glove' shaped palps, and two spines (spurs) on inner surfaces of front legs. Female larger than male, with larger abdomen and short spinnerets. **DISTRIBUTION** Restricted to Sydney basin NSW.

HABITAT AND HABITS Found in forests, woodland and urban areas, where it lies in wait near the end of its burrow for insects, including moths, and other spiders to come within striking distance. Burrow is within the ground, generally in open areas, and does not have a trapdoor over the entrance. Bite can be severely painful with local swelling, but only lasts a short period. However, due to this spider's close resemblance to the very dangerous Sydney Funnel-web (see p. 56), bites should be treated like bites from this species. Venomous

Sydney Funnel-web ▪ *Atrax robustus* CL 13.89mm

DESCRIPTION Glossed dark brown to black cephalothorax, and abdomen more brownish, duller and comparatively larger in female than in male. Spinnerets long, with terminal section the longest. Easily confused with the Sydney Brown Trapdoor (see p. 55), but male Sydney Funnel-web has smaller palps and a large spur on the second pair of front legs, used for protection from being attacked and killed by a female during mating. May also be confused with the Eastern Mouse Spider (see p. 54). **DISTRIBUTION** Central Coast (north) to Illawarra region (south), and west to Blue Mountains NSW. **HABITAT AND HABITS** Favours dark, moist areas in forest gullies, with male typically being more active than female, particularly in warmer months when seeking a mate and often entering gardens. Both sexes construct a long, silk-lined, tube-shaped burrow, normally in the ground, but it may also be in leaf litter. Male abandons his burrow after reaching maturity. Ground-dwelling species with large, powerful fangs that are able to pierce through toenails, although most people are bitten through the skin of the foot. Not all bites inject enough venom to cause severe reactions, but all bites should be treated as potentially lethal and medical attention for them should be sought immediately. Studies suggest that around 17 per cent of bites are capable of producing severe envenomation. Bites of males are more toxic than those of females. DANGEROUSLY VENOMOUS

Funnel-web spider ■ *Atrax yorkmainorum* CL 9.22mm

DESCRIPTION Glossed dark brown to black cephalothorax, and abdomen more brownish, duller and comparatively larger in female than in male. Spinnerets long, with terminal section the longest. Can be confused with the funnel-web *A. sutherlandi*, which may overlap in range in eastern NSW, and the Eastern Mouse Spider (see p. 54).

DISTRIBUTION South-eastern NSW and ACT. **HABITAT AND HABITS** Found in forests, where it favours dark, moist areas. Both sexes thought to construct a long, silk-lined, tube-shaped burrow, normally in the ground, but may also be in leaf litter. First described in 2010, and little is known about its breeding biology and habits. As is the case with other members of the genus, its bite should be regarded as potentially fatal; the venom of closely related species typically produces headaches, nausea, vomiting and abdominal pains. Venomous

Southern Tree Funnel-web ■ *Hadronyche cerberea* CL 10.72mm

DESCRIPTION Glossy black carapace, with large dark brown to blackish fang bases, and paler to dark brown abdomen. Males are more slender than females, with comparatively longer legs and a blunt spur on the second pair of legs. **DISTRIBUTION** South-eastern NSW as far north as Hunter River Valley NSW. **HABITAT AND HABITS** Found in open forests on rough-barked trees, where male shelters in cracks and holes, and under loose bark, and female creates a tubular silk shelter. Bites have caused severe envenomation in humans in about 75 per cent of recorded cases. Bites are capable of causing extreme pain, vomiting, profuse sweating, lung congestion and loss of consciousness, and should be regarded as potentially fatal, particularly in young children. DANGEROUSLY VENOMOUS

Northern Tree Funnel-web ■ *Hadronyche formidabilis* CL 14.84mm

DESCRIPTION Largest of all the funnel-webs. Both sexes very dark brown to black with large fang bases. Males more slender than females, with comparatively longer legs and a blunt spur on the second pair of legs.

DISTRIBUTION Southern Qld, south to around Hunter River Valley NSW. **HABITAT AND HABITS** Found in tall, undisturbed open forests and rainforests, where male shelters in cracks and holes, and under loose bark on trees, and female creates a tubular silk shelter. Bites have caused severe envenomation in humans in about 63 per cent of recorded cases. Bites are capable of causing extreme pain, vomiting, profuse sweating and loss of consciousness, and should be regarded as potentially fatal, particularly in young children. DANGEROUSLY VENOMOUS

Blue Mountains Funnel-web ■ *Hadronyche versuta* CL 11.18mm

DESCRIPTION Glossy black carapace, with dark brown to black legs. Male has smaller abdomen than female, with pale dorsal patch (female has reddish-brown patch).

DISTRIBUTION Found from Blue Mountains to Illawarra Region NSW (possibly as far south as Eden). **HABITAT AND HABITS** Occurs in forests and woodland, where it favours moist areas. Female constructs a burrow, generally in a rotting log or stump, with silken threads radiating from it and fragments of wood around the entrance. Males leave their burrows after their final moult to seek out a female. Bites are capable of causing extreme pain, vomiting, profuse sweating and loss of consciousness, and should be regarded as potentially fatal. DANGEROUSLY VENOMOUS

Barking Spider ■ *Selenocosmia stirlingi* TL 60mm

DESCRIPTION Large, hairy spider with no pattern on abdomen, large fangs, and long legs with thick, hairy feet. Spinnerets are long and finger-like. Males are smaller than females. **DISTRIBUTION** Throughout Australia, from eastern Qld, through inland NSW, north-western Vic, northern SA, southern NT and western WA. **HABITAT AND HABITS** Ground-dwelling spider of drier areas, where it hunts near the entrance of its burrow at night for insects, lizards, frogs and occasionally chicks of ground-nesting birds. Female maintains a silken lined burrow in moist ground, up to 1m deep, while males generally have a flimsy shelter under logs and rocks. Aggressive spider, capable of giving a painful bite and provoking severe allergic reactions, although most bites are reported to produce mild symptoms. Venomous

Giant Centipede ■ *Ethmostigmus rubripes* TL 16cm

DESCRIPTION Orange-yellow to dark blue-grey with black cross-bands on the 25 or 27 body segments. Legs yellow and arranged in 21 or 23 pairs, with first pair modified into large forcipules. **DISTRIBUTION** Throughout Australia. **HABITAT AND HABITS** Found in most habitats, wet and dry. Nocturnal, sheltering during the day in moist leaf litter, under logs, fallen bark or rocks, or in damp soil. Emerges on humid nights to hunt for insects, worms and snails, which are killed by a lethal injection of venom administered through the pincer-like forcipules. Venom is potent enough to kill quite large animals, including mice and lizards, and can cause intense pain that may last for several days, swelling and redness in humans. Venomous

Red-headed Centipede ■ *Scolopendra morsitans* TL 18cm

DESCRIPTION Mainly orange-yellow, with varying amounts of red on and behind the head, and with black cross-bands on the 25 or 27 body segments. Legs red, yellow or blue-grey, and arranged in 21 or 23 pairs, with first pair modified into large pincers. **DISTRIBUTION** Throughout Australia, and in many tropical countries around the world. **HABITAT AND HABITS** Found in most habitats, wet and dry. Nocturnal, sheltering

during the day in moist leaf litter, under logs or rocks, or beneath loose bark. Emerges at night to hunt for insects, spiders, other centipedes and small lizards, which are killed by a lethal injection of venom administered through the pincer-like forcipules. Venom can cause intense pain that can persist for several days, swelling and redness in humans. Venomous

Yellow Fever Mosquito ■ *Aedes aegypti* TL 3.8mm

DESCRIPTION Generally dark with white lyre-shaped pattern above, and white patches on abdomen, sometimes forming continuous bands. Legs have white bands. **DISTRIBUTION** Introduced into Australia and formerly widespread in WA, NT, Qld and NSW, but now considered restricted to northern and eastern Qld. Native to Africa. **HABITAT AND HABITS** Found around artificial water sources, including water tanks,

buckets, bird baths and underground pits, which are needed for the development of the aquatic young. Most active in shaded areas during the day and in the early evening. The only known vector of Dengue fever virus in Australia, but also capable of transmitting Ross River virus and Murray Valley encephalitis. Potentially Dangerous

Striped Mosquito ▪ *Aedes notoscriptus* TL 4mm

DESCRIPTION Dark greyish with conspicuous white to yellowish, lyre-shaped lines on dorsal shield, or scutum. Legs have pale bands, and proboscis has distinct white central band. **DISTRIBUTION** Throughout coastal and inland mainland Australia, including Tas. **HABITAT AND HABITS** Requires small pools of fresh water for larval development, with sites of these including tree hollows, garden ponds, pot plants, bird baths and gutters. Adults will travel up to 0.4km from the site where they developed. Readily attacks at dawn and dusk, but also in shaded areas during the day, and is a major pest in urban gardens. A known vector of Ross River virus and Barmah Forest virus, and able to carry Murray River encephalitis virus in laboratory tests. Potentially Dangerous

Before a blood meal   *After a blood meal*

Oriental Latrine Fly ▪ *Chrysomya megacephala* TL 11mm

DESCRIPTION Metallic blue-green body and large red eyes that meet in male, but are separated in female, and sponge-like mouthparts. Single pair of large, transparent wings. Second pair of wings has been modified into stabilizing appendages called halteres.

DISTRIBUTION Throughout Australasia and most of the world. In Australia, has been recorded in every state and territory except Tas. **HABITAT AND HABITS** Lives in close association with human habitation, and able to thrive in warm climates with increased breeding success, but also lives successfully in cooler areas. Adults attracted to human refuse, rotting carcasses and faeces, and have been recorded as carrying several types of disease-causing bacteria, including *Staphylococci typhi*, *Streptococcus aureus*, *Salmonella typhi*, *Pseudomonas aeruginosa* and *Bacillus* sp. Harmful

House Fly ▪ *Musca domestica* TL 8mm

DESCRIPTION Thorax and abdomen greyish, although abdomen is paler. Four blackish longitudinal lines on thorax, and single pair of transparent, triangular wings. Eyes are

reddish and mouthparts are sponge-like. **DISTRIBUTION** Throughout most of Australia and the rest of the world. **HABITAT AND HABITS** Abundant in urban and rural areas, wherever suitable breeding sites (human and animal waste) are found. Will investigate human foodstuffs, household rubbish, septic waste, and rotting vegetable and animal matter, using its sponge-like mouthparts to suck up liquids and to regurgitate saliva on to more solid foods to break them down. In doing so, it transfers microscopic organisms between these sites as it travels, which can lead to diseases such as salmonella, dysentery, hepatitis, cholera, poliomyelitis and typhoid fever. Harmful

Australian Bush Fly

▪ *Musca vetustissima* TL 7mm

DESCRIPTION Ashy-grey with two diverging black stripes on thorax, becoming four near head, and red eyes. The two rounded, triangular, transparent wings touch or overlap slightly when at rest. Abdomen yellowish and eyes touching in male; abdomen grey and black and eyes separated in female. **DISTRIBUTION** Endemic to Australia, where it is widespread in all states and territories except Tas. **HABITAT AND HABITS** Found in drier environments, where it is a bothersome fly during the summer months. Swarms of these flies land on the backs of animals, including humans, and crawl around the eyes, nose, ears and mouth, searching for moisture and proteins (female). Responsible for spreading bacteria such as *Escherichia coli* and *Salmonella*. Harmful

Stable Fly ■ *Stomoxys calcitrans* TL 7mm

DESCRIPTION Thorax and abdomen greyish. Darker longitudinal lines on thorax and pale area between these. The two transparent wings are held widely apart when not in flight. Piercing and sucking mouthparts. **DISTRIBUTION** Warmer parts of Australia, particularly where livestock industries are located. **HABITAT AND HABITS** Found mainly in rural areas and on beaches; seldom enters urban locations, except where livestock and horse stabling occur. Both sexes drink blood, and can consume up to three times their body weight. Bites are painful and can produce itchiness, urticaria and cellulitis. Allergic reactions can also occur in some people, accompanied by wheezing and hives. Harmful

March Fly ■ *Tabanus australicus* TL 13mm

DESCRIPTION Solidly built with large, reflective, iridescent eyes that meet in the middle in males. Eyes larger than those of the Stable Fly (see above), and green in female and reddish-brown in male. **DISTRIBUTION** Warmer parts of Australia, including Qld, NSW, Vic, WA and NT. **HABITAT AND HABITS** Favours moist forests and woodland, particularly in vicinity of water. Females drink blood to provide protein for their eggs to develop, and slice through a victim's skin with their sharp mouthparts. Males feed on nectar and other plant secretions. Bites are painful and can produce itchiness, lesions, urticaria, cellulitis and fever. Allergic reactions can also occur in some people, with symptoms including wheezing, hives, muscle weakness and potential anaphylaxis. Harmful

Biting midges ■ *Culicoides* spp. TL 3mm
(Sandflies)

DESCRIPTION Tiny, and the smallest of all flies that feed on blood. Adults resemble mosquitoes, with wings held in a V shape while feeding or at rest. **DISTRIBUTION**

Throughout Australia in suitable habitats. **HABITAT AND HABITS** Found mainly in coastal areas, including swamps, tidal flats, creeks, lagoons and mangroves, where they are most active at dawn and dusk. Mainly feeds on nectar and other plant exudates, but females readily attack the exposed skin of humans to obtain blood. The victim often is not aware of being bitten until swelling, itchiness and redness appear. Sensitive people can exhibit severe local allergic reactions, including sores that persists and weep for weeks. Secondary infection can also result from scratching at the itchy sores. Harmful

Rat Flea ■ *Xenopsylla cheopis* TL 4mm

DESCRIPTION Dark reddish-brownish, with long, powerful hindlegs, laterally compressed, wingless body and hard exoskeleton. Mouthparts adapted for piercing and sucking.

DISTRIBUTION Worldwide in tropical, subtropical and some temperate regions. **HABITAT AND HABITS** Widespread in urban areas and, in homes, found in clothing and bedding near sleeping areas of human hosts. Piercing and sucking mouthparts are used to feed on the blood of host animals, normally rats, but murine typhus, or rat-flea typhus, can be transmitted to humans when a flea's droppings enter the bloodstream. Symptoms include shaking, chills, headache, fever, rash and potential mortality in elderly people. This species is a vector for the bacterium *Yersinia pestis*, which was transferred from rats to humans and caused the famous bubonic plague of the Middle Ages. Harmful

Human Body Louse

■ *Pediculus humanus corporis* TL 4mm

DESCRIPTION Wingless insect with large abdomen and slightly narrower head. Legs have sharp claws for gripping hair or clothing of host, and piercing and sucking mouthparts. **DISTRIBUTION** Worldwide, wherever humans congregate. **HABITAT AND HABITS** Lives on clothing of humans who live in crowded conditions with poor hygiene, and spread by close contact with other humans in the same area. Feeds on blood, and bites from it can result in skin reactions including rashes and itchiness. Known to spread diseases such as epidemic typhus, trench fever and louse-borne relapsing fever. Body lice do not persist in clean environments where regular bathing and laundering of clothes and bed linen are practised. Harmful

Honey Bee ■ *Apis mellifera* Size TL 16mm

DESCRIPTION Generally yellowish-brown all over, with blackish-brown bands on abdomen and orange-brown wash. Some variation in colour and pattern between individuals. **DISTRIBUTION** Introduced into Australia by early European settlers and now widespread. **HABITAT AND HABITS** Occurs in forests, woodland, heaths and urban areas, wherever nectar-producing flowering plants are found. Lives in a large communal nest, or hive. The sting, which it uses to protect the colony, is barbed and stays embedded in its victim when it flies away (the bee dying as a result), while the venom gland continues to pump the venom into its victim. On average about 1,000 people are hospitalized in Australia each year due to anaphylaxis to bee venom. Venomous

Swarming bee colony

Foraging worker

European Wasp ■ *Vespula germanica* TL 1.5cm
(Yellowjacket)

DESCRIPTION Black with broad yellow bands and almost hairless, elongated abdomen.
DISTRIBUTION Introduced into southern Australia, where it is widespread from southern

Qld in east to Perth WA in west, and in Tas. Native to Europe, North Africa and Asia.
HABITAT AND HABITS Found in relatively cool, moist areas, where it builds a papery nest in a crevice or tree hollow, generally close to the ground, but nests have also been found in roof spaces of dwellings. Food consists of insects, carrion and sugary substances, including fruits and honeydew, as well as manufactured cakes and drinks. Readily crawls inside soft drink cans and bottles to access the contents. Humans are at risk of being stung in the mouth when they drink from a container with a wasp inside. Stings anywhere on the body cause pain, inflammation and itchiness, and allergic reactions occur in some people. Potentially Dangerous

Paper wasps ■ *Polistes* spp. TL 2.2cm

DESCRIPTION Predominantly black with two pairs of brownish wings, and scattered yellow or orange markings or rings on long, cylindrical abdomen. Very narrow waist

and small head with large eyes.
DISTRIBUTION Throughout Australia. **HABITAT AND HABITS** Occurs in variety of habitats throughout its range, including urban areas, forests and woodland. Forms small colonies that build distinctive papery nests from a mixture of wood shavings and saliva. Adults aggressively defend their nest. Sting injects venom that causes intense pain and swelling around the immediate area, and a victim can be stung multiple times by the same wasp. Venom can cause allergic reactions and anaphylaxis in some people. Potentially Dangerous

Tarantula Hawk ■ *Cryptocheilus bicolor* TL 35mm

DESCRIPTION Blackish body with broad, dark yellow-orange tip on abdomen, orange-brown wings (without black tips), yellow-orange head and long yellow antennae.

DISTRIBUTION Throughout Australia in suitable habitats.

HABITAT AND HABITS Found in forests, woodland, wetlands, heaths and adjacent urban areas, where adults feed on nectar and fruits. Female hunts for large huntsman and wolf spiders as food for its larvae after they hatch. She paralyses her prey and drags it to a burrow before laying a single egg on it. Neither the male nor the female is aggressive towards humans, and these wasps rarely sting. In cases when they have done so, however, the sting has been described as excruciatingly painful. Venomous

Blue Ant ■ *Diamma bicolor* TL 3cm

DESCRIPTION Female has metallic blue or greenish body and head, and orange-brown legs and antennae, and is wingless. Male winged and blackish, with incomplete whitish bands on abdomen.

DISTRIBUTION Throughout south-eastern Australia in suitable habitats, from south-eastern Qld, through eastern NSW and Vic, to Tas. **HABITAT AND HABITS** Found in forests, woodland and urban parks and gardens, where it feeds mostly on nectar. Flying male picks up female and mates with her in flight, after which she constructs a burrow in which to lay her eggs. Sting of female can cause intense pain, swelling and allergic reactions in humans. Venomous

Bottlebrush Sawfly ■ *Pterygophorus cinctus* TL 4cm

DESCRIPTION Larvae mostly hairless, with blackish-brown head that becomes paler and reddish with age, and stout reddish, orange, greenish and yellowish body ending in long, narrow dark tip. **DISTRIBUTION** Throughout south-eastern Australia in suitable habitats, from south-eastern Qld, through eastern NSW, to southern Vic and Tas. **HABITAT AND HABITS** Found in habitats in which bottlebrushes grow, where female lays her eggs inside the stems and leaves. After hatching, the young larvae are gregarious, feeding together and skeletonizing the leaves of the tree they were laid on, but feeding on separate leaves when they grow larger. Once fully grown, they drop to the leaf litter and pupate without a cocoon. If handled, the larvae may secrete an irritating liquid. Harmful

Adult   *Larvae*

Giant Bull Ant ■ *Myrmecia brevinoda* TL 26mm
(Giant Bulldog Ant)

DESCRIPTION Variable, but generally reddish-brown, with black abdomen, large eyes and

powerful, forwards projecting jaws. Typically has a long body, and is among the longest ants in the world. **DISTRIBUTION** Eastern Qld, NSW and eastern Vic. **HABITAT AND HABITS** Found in rainforests, open forests and woodland, where it typically nests in soil or under logs and rocks. Abdomen has a stinger that is used to envenom and subdue insect prey, but also feeds on honeydew and other sweet secretions from plants and insects. Sting is painful, and venom has caused severe allergic reactions and fatalities in humans. Venomous

Jumper Ant ▪ *Myrmecia nigrocincta* TL 17mm
(Jack Jumper)

DESCRIPTION Mainly black and orange to reddish-brown, with large, yellowish, forwards pointing jaws and large eyes. **DISTRIBUTION** Coastal eastern Australia, from northern Qld to eastern Vic. **HABITAT AND HABITS** Most often found in wet forests and rainforest fringes, where it nests underground at the base of a grass tussock or tree. Workers forage for other insects and honeydew, and can be aggressive, particularly when defending their nest and territory. Capable of stinging humans, jumping forwards and gripping the skin with its jaws while stinging with its abdomen, often doing so several times in quick succession. Venom causes intense localized burning pain and swelling. It is capable of producing severe allergic reactions and has been responsible for human fatalities. Venomous

Green-head Ant
▪ *Rhytidoponera metallica* TL 7mm

DESCRIPTION Distinctive metallic green or purplish, with long, segmented antennae. **DISTRIBUTION** Endemic to subtropical Australia and found in every state except Tas. Introduced into NZ. **HABITAT AND HABITS** Occurs in variety of habitats, including forests, woodland, deserts, and urban parks and gardens, and recovers quickly after effects of bushfires. Nests can be in stumps or underground, below fallen debris, with colonies feeding during the day on variety of invertebrates and honeydew. Aggressive and can give a very painful sting. Venom can cause severe allergic reactions and anaphylaxis in some people, and can be fatal. Venomous

Red Fire Ant ▪ *Solenopsis invicta* TL 6mm

DESCRIPTION Coppery-brown, darker on the abdomen, and antennae with distinctive two-segmented, clubbed terminal ends. Members of a single colony vary greatly in size. **DISTRIBUTION** Introduced from South America, and recorded in central and south-east Qld, and Sydney region NSW. **HABITAT AND HABITS** Found in variety of open

disturbed habitats in hotter environments. Can form large colonies of up to 400,000 individuals. Feeds on invertebrates, vertebrates (including carrion) and plants. Aggressive and can give a very painful sting, with a burning sensation that can last up to an hour. Itchy pustules that can persist for up to 10 days may also form around bite site. Can cause severe allergic reactions and anaphylaxis in some people. Venomous

American Cockroach ▪ *Periplaneta americana* TL 55mm

DESCRIPTION Large, reddish-brown cockroach, with pale yellowish-brown margin on pronotal shield. Large wings (extending past tip of abdomen in males) and a competent flier. **DISTRIBUTION** Native to Africa and introduced to most countries around the world, including Australia. **HABITAT AND HABITS** Occurs in close association with urban human habitation. Mostly found indoors, where it is active at night. Pest species of urban households and similar dwellings, where it favours darker places during the day, including subfloor areas, roof voids, wall cavities and gardens. Eats most organic matter, particularly fermenting foods, and is a vector for a number of bacterial diseases and a leading cause of allergic reactions. Harmful

Australian Cockroach ▪ *Periplaneta australasiae* TL 35mm

DESCRIPTION Brown, with yellow border around pronotum and longitudinal yellow stripes on outer edges of wings. Very similar to the larger American Cockroach (see opposite). **DISTRIBUTION** Introduced, and now widespread in tropical and subtropical areas of Australia. **HABITAT AND HABITS** Occurs in close proximity to human habitation. Found in urban areas, where it is common in gardens and outside areas, but will enter houses to find food and is a competent flier. Pest species of households and similar dwellings, where it feeds on a variety of vegetable matter but also scavenges on other organic matter. Known to spread disease-causing bacteria. Harmful

Adult

Young

German Cockroach
▪ *Blattella germanica* TL 15mm

DESCRIPTION Pale brown to dark brown, with two dark longitudinal stripes on pronotum, which distinguish it from all other similarly sized species. **DISTRIBUTION** Native to Africa and introduced to Australia. **HABITAT AND HABITS** Closely associated with human habitation. Most often found around food-preparation and storage areas in houses and commercial buildings, where it is largely nocturnal. Female carries egg case, or ootheca, within her body. Identified in spread of a number of diseases, including salmonella and typhoid, and over 170 bacterial isolates have been recorded on its body. Harmful

Black Slug Moth ▪ *Doratifera casta* WS 40mm (adult); TL up to 30mm (caterpillar)

DESCRIPTION Adult moths have brown wings, with two black dots on forewings. Caterpillars black, with white blotches and numerous off-white, fleshy spikes, four of which

produce tufts of stinging hairs. Legs are short. **DISTRIBUTION** Qld, NSW and Vic. **HABITAT AND HABITS** Found in wooded areas where its favoured food plants grow, mainly eucalypts, acacias and bottlebrushes. Female lays batches of around 40 brown, hairy eggs on the same leaf. After hatching, caterpillars initially feed together on the same leaf, but move to separate leaves as they grow in size. Stinging hairs of both adults and caterpillars are capable of causing minor to moderate skin irritations that can remain itchy for several days. Harmful

Painted Cup Moth ▪ *Doratifera oxleyi* WS 50mm (adult); TL 35mm (caterpillar)

DESCRIPTION Adult female mottled orange and brown, while male has orange and white-tipped abdomen. Caterpillar pale greyish-green and white, with numerous fleshy

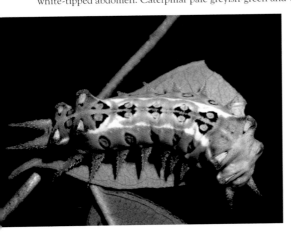

spikes, and four pairs of tufted clear to yellowish stinging hairs. **DISTRIBUTION** South-eastern mainland Australia, from central eastern NSW, through southern Vic and Tas, to southern SA. **HABITAT AND HABITS** Found in eucalypt forests, woodland and adjacent urban parkland, where female lays batch of eggs on leaves of gum trees, covering the cluster with stinging hairs from her body. Stinging hairs of both adults and caterpillars are capable of causing an itchy rash that can persist for several days. Harmful

Mottled Cup Moth ▪ *Doratifera vulnerans* TL 20mm (caterpillar)

DESCRIPTION Adult mottled brown and hairy. Caterpillar pale greyish-green to pinkish-brown, with colourful yellow, green and black saddle across middle of body, and four pairs of tufted greyish-brown stinging hairs that are extended when disturbed. **DISTRIBUTION** Eastern and south-eastern mainland Australia, from central eastern Qld to southern Vic. **HABITAT AND HABITS** Inhabits eucalypt forests, woodland and adjacent urban parkland, where female moth lays her eggs on leaves of gum trees, upon which young caterpillars feed in groups after hatching. Stinging hairs of both adults and caterpillars are capable of causing minor to moderate skin irritations that can remain itchy for several days. Harmful

Common Anthelid ▪ *Anthela acuta* TL 50mm (caterpillar)

DESCRIPTION Caterpillar brown and hairy, with pairs of white-and-pink spots, and longer tufts of pale brown to whitish hairs. **DISTRIBUTION** Eastern Australia, with most caterpillars collected in south-eastern Qld, eastern NSW, Vic and Tas. **HABITAT AND HABITS** Found in wide range of habitats, where caterpillars feed mainly at night on variety of vegetation, depending on the race and distribution, including grasses, palms, wattles, English Oak *Quercus robur* and European Beech *Fagus sylvatica*. When fully grown the caterpillar constructs a hairy cocoon within an outer wall of leaves, placed on the ground in debris. Its long hairs can cause mechanical skin irritations. Harmful

Urticating Anthelid
■ *Anthela nicothoe* TL 80mm (caterpillar)

DESCRIPTION Caterpillar brown and hairy, with body segments having a small hairless gap separating them, and a pinkish inverted 'Y' on front of head. Female larger than male (50mm). **DISTRIBUTION** Eastern Australia, including south-eastern Qld, eastern NSW, Vic, Tas and south-eastern SA. **HABITAT AND HABITS** Found in wooded habitats, where caterpillars feed on wattles, particularly the Cootamundra Wattle *Acacia baileyana* and Silver Wattle *A. dealbata*. Long hairs of caterpillar and its cocoon can easily puncture human skin and break off, causing mechanical skin irritations (urticaria) in some people. Harmful

White-stemmed Wattle Moth
■ *Chelepteryx chalepteryx* TL 70mm (caterpillar)

DESCRIPTION Caterpillar reddish-brown, with conspicuous paler line along dorsal surface and numerous raised paler spots, each containing a tuft of short bristles. **DISTRIBUTION** Eastern Australia, including Qld, NSW and Vic. **HABITAT AND HABITS** Found in forests, woodland and heaths, and adjacent urban areas, where caterpillar feeds on leaves of several different plants, including wattles and lilies. When fully grown, caterpillar builds cocoon in a tree crevice or under bark, which is covered with bristles. Bristles of caterpillar can cause pain and skin irritations (urticaria). Harmful

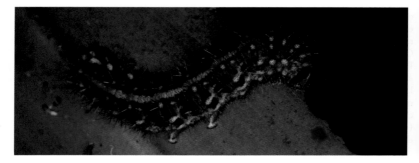

White-stemmed Gum Moth ■ *Chelepteryx collesi* TL 120mm (caterpillar)

DESCRIPTION Caterpillar strongly banded white and reddish-brown to blackish, with numerous raised yellow or reddish-brown spots, each with tufts of shorter reddish-brown or longer white bristles.

DISTRIBUTION Much of south-eastern Australia, from southern Qld, through NSW and ACT, to Vic.

HABITAT AND HABITS Found in forests, woodland, heaths and urban areas, where caterpillars are known to feed on leaves of various eucalypts, angophoras and the Brush Box *Lophostemon confertus*. Once fully grown, caterpillar builds cocoon with two walls of silk, covered in numerous short bristles. Bristles of caterpillar and cocoon are barbed and cause intense pain and skin irritations (urticaria). Harmful

Perfect Tussock Moth ■ *Calliteara pura* TL 40mm (caterpillar)

DESCRIPTION Caterpillar yellowish to orange and white, with numerous long hairs that change from yellowish to white after moulting. **DISTRIBUTION** Qld, NSW and Vic.

HABITAT AND HABITS Common garden pest, but also found in woodland and other bushland. Female lays up to 200 small, rounded eggs, and after hatching the caterpillars feed on leaves of a variety of native and exotic plants, including roses, magnolias, and the Gymea Lily and Bird of Paradise. Once fully grown, each caterpillar builds a large cocoon of hairs and moulted skins. The long hairs can cause skin irritations (urticaria), and are capable of drifting over long distances if dislodged. Harmful

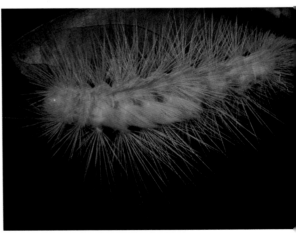

Vampire Moth
▪ *Calyptra minuticornis* WS 50mm

DESCRIPTION Moth generally pale reddish-brown with a silverish sheen, and leaf-like, with diagonal lines that mimic veins in a leaf. **DISTRIBUTION** In Australia, occurs in eastern Qld and eastern NSW. More widely distributed in PNG, Indonesia, Malaysia, Taiwan and Japan. **HABITAT AND HABITS** Found mainly in rainforests, where the moth normally feeds on fruit juices, but will also feed on the blood of animals, including humans (in laboratory tests), and symptoms can include intense pain and swelling that can persist for up to 24 hours. Harmful

Mistletoe Browntail Moth ▪ *Euproctis edwardsii* TL 20mm (caterpillar)

DESCRIPTION Caterpillar greyish-black, with paler white lines on dorsal surface, red spots along sides and long white hairs. **DISTRIBUTION** South-eastern Australia. **HABITAT AND HABITS** Found in eucalypt woodland and adjacent areas, where caterpillars feed on mistletoe. Long, barbed hairs of caterpillars and their discarded skins break off with contact and are difficult to remove. They can cause urticarial wheals and papular eruptions. They can also drift large distances in the wind and become troublesome to sensitive people over a wide area. Harmful

Yellow Tussock Moth

◾ *Euproctis lutea* WS 30mm (adult); TL 15mm (caterpillar)

DESCRIPTION Caterpillar black and hairy, with white spot on thorax and white longitudinal line along upper surface of abdomen. Moth generally yellow with paler wavy lines across forewings. **DISTRIBUTION** Northern Australia, including WA, NT, Qld and NSW. **HABITAT AND HABITS** Found in variety of wooded habitats and urban parks and gardens, where it feeds on leaves of freshwater mangrove, cocky apple and cultivated plants, including tomatoes and roses. Contact with the hairs on a caterpillar, egg, adult moth and even the site where caterpillars have been feeding can cause urticaria, which can become extremely itchy and swollen in sensitive individuals. More severe reactions can appear, especially in young people, and medical attention may be required. Harmful

Patterned Notodontid

◾ *Aglaosoma variegata* WS 30mm (adult); TL 40mm (caterpillar)

DESCRIPTION Caterpillar generally pale grey above, with yellowish-brown raised lumps along dorsal surface, and rows of blue elongated spots and reddish-brown mottling along sides. Numerous pale-tipped brown to black hairs, coronets of shorter stiffer hairs and hairy head. Adult moths have patterned black-and-white forewings, with scattered small red spots, greyish hindwings and banded black-and-orange abdomen. **DISTRIBUTION** Eastern Australia from Atherton Tableland Qld, through NSW to Vic, possibly as far west as eastern SA. **HABITAT AND HABITS** Found in range of habitats, where caterpillars feed on variety of plants, including wattles, banksias, she-oaks and rushes. Hairs of both moths and caterpillars can cause severe skin rashes and infection. Harmful

Common Epicoma
■ *Epicoma melanosticta* WS 30mm (adult); TL 40mm (caterpillar)

DESCRIPTION Caterpillar blackish, with long brownish and grey hairs, and red band around rear of yellow-and-black head. **DISTRIBUTION** Throughout southern Australia, with records in each state and territory except NT. **HABITAT AND HABITS** Found

in variety of wooded habitats, where food plants grow. Female lays eggs in hairy mass of hairs, and after hatching caterpillars initially feed in groups on myrtles and tea trees, but feed individually when they are larger. Long hairs of caterpillars and moths are capable of causing urticaria. Harmful

Mottled Epicoma
■ *Epicoma protrahens* WS 30mm (adult); TL 40mm (caterpillar)

DESCRIPTION Caterpillar yellow to green, with two conspicuous black, oval-shaped spots on first two segments of abdomen, and dark brown and orange spots on remaining

segments. Body sparsely covered with long white hairs, and head orange-brown. **DISTRIBUTION** From Atherton Tablelands Qld south to at least Batemans Bay NSW. **HABITAT AND HABITS** Found in eucalypt forests. After hatching, caterpillars initially feed in groups on bottlebrushes and paperbarks, but feed individually when they are larger. Long hairs of caterpillars and moths are capable of causing urticaria. Harmful

Processionary Caterpillar

■ *Ochrogaster lunifer* WS 30mm (adult); TL 40mm (caterpillar)

DESCRIPTION Caterpillar grey, with long white hairs and brownish head. **DISTRIBUTION** Suitable habitat throughout mainland Australia. **HABITAT AND HABITS** Found in wooded habitats, where caterpillars are well known for forming a long processional train as they walk head to tail in a line. They are voracious nocturnal feeders, and large groups are capable of stripping the foliage completely from a tree before moving to another. During the day they huddle in a mass at the base of the food tree. Long hairs of caterpillars and moths are capable of causing urticaria. Harmful

Gum-leaf Skeletoniser ■ *Uraba lugens* TL 20mm (caterpillar)

DESCRIPTION Caterpillar brown with longitudinal lines of large and small yellow spots along back and sides, and numerous long white hairs. Head has a long, ridged crown, formed from the dried skins of past moults. **DISTRIBUTION** Throughout most of southern and eastern Australia, from north-eastern Qld, to south-western WA and Tas. **HABITAT AND HABITS** Occurs in forests and woodland, where young caterpillars feed together on leaves of eucalypts and the Brush Box *Lophostemon confertus*, forming a long row that moves down a leaf, and leaving only the leaf veins. Hairs of caterpillar can cause pain, redness and urticarial weals. Harmful

Bee Killer Assassin Bug ■ *Pristhesancus plagipennis* TL 25mm

DESCRIPTION Yellowish to orange-brown and dark grey as adults, with transparent wings, long head, large eyes and thickened curved rostrum (proboscis). **DISTRIBUTION** Mainly in coastal eastern Australia, from northern Qld to around Sydney NSW. **HABITAT**

AND HABITS Occurs in open forests and woodland, and similarly vegetated areas, including urban parks, gardens and cropland. Hunts among foliage for soft-bodied insects, particularly the Honey Bee (see p. 65), caterpillars and spiders, which it grasps with its front legs and stabs with its powerful rostrum before sucking out the victim's body fluids. Capable of inflicting an intensely painful bite and potential disease in humans. Harmful

Common Bed Bug ■ *Cimex lectularius* TL 5mm

DESCRIPTION Orange-brown to reddish-brown, with oval-shaped, wingless body. When not feeding, proboscis is held flat against underside of thorax. **DISTRIBUTION** Found in close association with dense populations of humans, particularly in capital cities, and readily transported between locations. **HABITAT AND HABITS** Occurs in small groups in

sleeping areas of its host. Spends most of the day and night sheltering in a dark crevice, emerging for a few hours before dawn to feed on the host's blood. Female needs blood for egg production, and eggs are laid in small batches in crevices near the sleeping area of the host. While this species has not been responsible for major disease transmission, severe bed bug infestations can lead to iron deficiency in humans. Harmful

Crown-of-thorns Starfish ■ *Acanthaster planci* W 35cm

DESCRIPTION Variably coloured, but generally orange-red to purple, with central disc and up to 23 arms. Upper surface covered in long spines. **DISTRIBUTION** Within Australian waters, common on GBR and around Lord Howe Island. More generally, in marine waters throughout Indo-Pacific. **HABITAT AND HABITS** Found on coral reefs, particularly in sheltered, relatively shallow areas around 10m deep, where it feeds almost exclusively on hard corals by pushing out the gastric folds of its stomach through its mouth (eversion). Each arm has a light-sensitive eye at the tip, which it uses to navigate through its surroundings. Long spines are venomous, and a sting can cause intense pain, swelling, redness, vomiting and potential anaphylaxis in humans. Although rare, at least one human fatality has been recorded. Venomous

Black Longspine Urchin ■ *Diadema setosum* Test 80mm

DESCRIPTION Usually black, but some spines can be paler or banded. Spherical test and numerous very long, flexible spines above and shorter spines below. Distinguished by five bluish spots (iridophores) on test. **DISTRIBUTION** Within Australia, from central coast of WA, through tropical north around NT and Qld, to around Sydney NSW. More generally, in marine waters throughout tropical Indo-Pacific. **HABITAT AND HABITS** Found in rocky areas of intertidal zone and coral reefs to depths of around 30m, where it grazes on algae. Spines are barbed and brittle, and can easily pierce human skin and break off. Venom within them can cause localized pain, swelling and redness – although the symptoms subside after a few hours, embedded fragments can cause ongoing irritation and potential infections. Venomous

Black-spined Sea Urchin ■ *Diadema savignyi* Test 80mm

DESCRIPTION Resembles confusingly similarly named Black Longspine Urchin (see p. 81), which has longer spines. **DISTRIBUTION** Widespread in Indo-Pacific, including

Northern Australia. **HABITAT AND HABITS** Found in shallow inshore waters to around 70m depth, where it inhabits the sandy sea floor and coral reefs of the continental shelf. Often seen sheltering in rocky crevices during the day, and grazing at night on algae, seagrass and some invertebrates. Venomous spines are barbed and brittle, and easily pierce human skin and break off, causing short-term localized pain, swelling and redness, and embedded fragments can cause ongoing problems. Venomous

Flower Urchin ■ *Toxopneustes pileolus* Test 13.5cm

DESCRIPTION Test roughly circular and flattened on oral surface, with five segments, and variegated pattern of grey, red and purple. It has numerous large, pincer-like organs, called pedicellariae, among the tubular feet, which are yellowish-white to pinkish-white. **DISTRIBUTION** In Australia, occurs in north and east, from northern WA, through NT and Qld, and south to around Montague Island NSW. Generally widespread in marine waters of Indo-West Pacific. **HABITAT AND HABITS** Found on coral reefs and in areas with sandy or rocky substrates, where it feeds mainly on algae and aquatic invertebrates. Larger pedicellariae are venomous with fang-like tips, and are used for defence. Venom can cause intense pain in humans, accompanied by lightheadedness, fainting, paralysis and respiratory distress; fatalities have also been recorded. Venomous

The test is roughly circular

Close up showing 'pincer-like' pedicellariae

Broadnose Shark ■ *Notorynchus cepedianus* TL 3m

DESCRIPTION Grey to brownish above and white below, with seven tall gill slits located forwards of pectoral fin. Single small dorsal fin and blunt snout. **DISTRIBUTION** In Australian waters, found from central NSW south to Tas, and west to south-western WA. **HABITAT AND HABITS** Occurs in temperate marine waters to around 200m depth, where it is mostly found close to the sea bottom, although it has been recorded near the surface, particularly in shallower waters. Although rarely encountered, it has been responsible for at least three unprovoked non-fatal attacks on scuba divers, surfers and beachgoers. The most recent attack occurred when a surfer was knocked off her board, and did not result in any injury. Potentially Dangerous

Crested Hornshark ■ *Heterodontus galeatus* TL 1.5m

DESCRIPTION Generally grey to brown above, with large blackish blotches, and white below. Two large head crests, each ending sharply at rear of skull. **DISTRIBUTION** Confined to east coast of Australia, from southern Qld to southern NSW. **HABITAT AND HABITS** Found in inshore marine waters, where it inhabits the seafloor at depths of up to 100m, resting on the sand, in rocky crevices or in seagrass beds, and feeding on oysters, fish, crustaceans and echinoderms. Both the male and female have a blunt spine at the front of each of the two dorsal fins, sharper in juveniles, which is presumed to be venomous. This docile species is generally regarded as harmless to humans, but can inflict a painful bite if provoked. Harmful

Port Jackson Shark ■ *Heterodontus portusjacksoni* TL 1.7m

DESCRIPTION Grey to brown above, with distinctive, harness-like blackish lines along sides of body, along top of head and down sides of head. Head has a raised longitudinal crest above each eye.

DISTRIBUTION Coastal and offshore subtropical waters of Australia, from southern Qld, through NSW, Vic, Tas and SA, to central WA. **HABITAT AND HABITS** Found on coastal reefs, mainly on the continental shelf, where it inhabits the seafloor at depths of up to 275m, normally in rocky areas, but also rests on the sand or in seagrass beds. Feeds on oysters, fish, crustaceans and sea urchins. Both the male and female have a blunt spine at the front of each of the two dorsal fins, sharper in juveniles, which is regarded as venomous. Docile and generally considered harmless to humans, but can inflict a painful bite if provoked. Harmful

Grey Nurse Shark ■ *Carcharias taurus* TL 3.6m

DESCRIPTION Grey to bronze-brown above, with two similarly sized dorsal fins, and whitish below. Young often spotted with reddish-brown on rear of body and along tail.

DISTRIBUTION Tropical and temperate waters of all states in Australia except Tas. Wider roaming throughout Indian, Western Pacific and Atlantic Oceans. **HABITAT AND HABITS** Most commonly seen in shallow inshore waters, but can occur at depths of 190m or more. Feeds on fish, which are detected using electroreceptors located on the lower jaw. Mouth is filled with numerous sharp, fang-like teeth, which are constantly replaced as they are lost. Although this species is generally considered harmless, bites from it have been recorded. Diving with Grey Nurse sharks is popular at many sites in NSW and Qld. Harmful

Great White Shark ■ *Carcharodon carcharias* TL 5m
(White Pointer)

DESCRIPTION Pale to dark grey above and white below except for black tips on pectoral fins. Cone-shaped snout, black eyes and large, triangular dorsal fin. **DISTRIBUTION** All Australian waters, but most common in south. Wide roaming in most of the world's seas and oceans. **HABITAT AND HABITS** Oceanic, and particularly associated with the pelagic zone of continental shelf waters, although juveniles may enter large lakes and estuaries. Skilled hunter of fish, cephalopods, crustaceans, seabirds and marine mammals, but also feeds on carrion. In the three years 2013–2015, responsible for 28 unprovoked attacks in Australia, with two proving fatal. As a comparison, during the same three-year period around 3,500 people were killed in motor vehicle-related accidents. DANGEROUS

Shortfin Mako ■ *Isurus oxyrinchus* TL 4m

DESCRIPTION Slender, streamlined body and pointed snout. Blue above, becoming paler on sides, and white on underparts. Teeth long, slender and smooth edged, and protrude out of the mouth. Eyes large and dark. **DISTRIBUTION** All coastal Australian tropical and temperate marine waters. Found throughout the world,. **HABITAT AND HABITS** Oceanic, mainly occuring in waters above 16° C. Feeds on fish, squid and marine mammals, which are taken either close to the surface or at depths down to around 150m. Very fast-swimming shark and possibly the fastest of all shark species. Unprovoked attacks on humans are extremely rare, but injuries to anglers occur due to the shark's spectacular leaps out of the water when hooked and landed on to watercraft. DANGEROUS

Common Thresher Shark ■ *Alopias vulpinus* TL 5.5m

DESCRIPTION Blue-grey above with very long upper lobe on caudal fin. White of underside extends above pointed pectoral fins. **DISTRIBUTION** In Australia, absent only

from northern WA, NT, Gulf of Carpentaria and northern Qld. Occurs in tropical to temperate marine waters worldwide. **HABITAT AND HABITS** Found in marine waters at various depths from shallow coastal waters to around 370m. Feeds on small fish, which it herds and stuns with its large tail. Shy and generally hard to approach, this species is likely to pose minimal danger to humans, but some divers have reported being hit by the powerful tail and attacks on boats have occured. Harmful

Banded Carpet Shark ■ *Orectolobus halei* TL 2m

DESCRIPTION Ornately coloured, dorsally flattened shark. Generally pale brown above, with large darker brown blotches edged with black, each containing smaller, black-edged paler blotches. **DISTRIBUTION** Waters of southern Australia, from around Port

Hedland WA to southern Qld. **HABITAT AND HABITS** Bottom-dwelling inhabitant of coastal and inshore waters, including bays and lagoons, up to depths of 100m, where it hunts around coral reefs and offshore islands for fish, crabs, crayfish and octopuses. Although it is generally docile, its large mouth contains many long, pointed teeth that are capable of inflicting a nasty wound, presumably in self-defence or due to mistaken identity. Regularly encountered by divers and has been known to steal fish from fishermen. Harmful

Spotted Wobbegong ■ *Orectolobus maculatus* TL 3m

DESCRIPTION Ornately coloured, dorsally flattened shark with distinctive skin flaps on snout. Yellowish-brown to greenish above, with large darker brown blotches (saddles), and numerous white spots and rings. **DISTRIBUTION** Coastal and inshore waters from around Broome WA in west, south through SA, Vic, Tas and NSW, to islands off central Qld. **HABITAT AND HABITS** Bottom-dwelling inhabitant of shallow coastal and inshore waters to depths of around 100m, where it hunts around coral reefs for fish, crabs, crayfish and octopuses. Although it is generally docile, its large mouth and long, pointed teeth are capable of inflicting a nasty wound, and several attacks on swimmers, surfers and other beachgoers have been reported, presumably mostly in self-defence after the shark has been stepped on or provoked. Harmful

Banded Wobbegong
■ *Orectolobus ornatus* TL 1.1m

DESCRIPTION Yellowish-brown above with large patches of darker brown, edged darker, and numerous pale blue-grey spots, and yellowish-olive below. Head broad and flat with small eyes, and numerous skin flaps along snout. **DISTRIBUTION** East Australian waters, from northern Qld to southern NSW. **HABITAT AND HABITS** Bottom-dwelling shark of tropical to warm temperate coastal and inshore waters, often around reefs, to depths of around 50m. Generally docile and readily approached by divers, but has been known to attack if provoked, and is capable of causing a painful wound and potential infection. Beachgoers have also been attacked, presumably when a shark has been stepped on or due to a case of mistaken identity. Overfishing has reduced numbers of this species. Harmful

Silvertip Shark ■ *Carcharhinus albimarginatus* TL 3m

DESCRIPTION Grey to brownish above and pale below, with an obscure paler longitudinal stripe along sides, and long, rounded snout. White tips on main dorsal fin, and on upper lobe of caudal fin and trailing margins of all major fins. **DISTRIBUTION** Northern Australia, from around Hervey Bay Qld, to Shark Bay WA, but seemingly absent from Gulf of Carpentaria and Arafura Sea. Generally patchy distribution in tropical Pacific

and Indian Oceans. **HABITAT AND HABITS** Inhabits offshore waters around islands, reefs and off the continental shelf, from surface waters to about 600–800m depth. Taken as a bycatch of longline fishing and trawling, and killed for its meat, fins, organs and cartilage. Aggressive, fast-moving fish that is capable of causing severe injuries to humans, but is rarely encountered. Potentially Dangerous

Grey Reef Shark ■ *Carcharhinus amblyrhynchos* TL 1.8m

DESCRIPTION Bronze to grey above with triangular main dorsal fin, occasionally with a small white tip. Large pale eyes with dark pupils. Indistinct pale greyish-white stripe behind pelvic fins and along tail, and caudal fin edged with black along rear edge. **DISTRIBUTION** Tropical marine waters of northern Australia, from around Port Hedland WA to southern Qld. **HABITAT AND HABITS** Occurs in inshore and offshore waters to

around 280m depth, often around coral reefs, where it feeds mainly at night, predominantly on fish, but also on cephalopods and some crustaceans. Inquisitive and will readily approach divers, and capable of causing major injuries, particularly if agitated. When threatened, raises its head, arches its back and swims with exaggerated movements – if this occurs, people in the vicinity should exit the water immediately. Potentially Dangerous

Bronze Whaler ■ *Carcharhinus brachyurus* TL 3m

DESCRIPTION Greyish-brown to bronze above and whitish below, with an obscure paler stripe along sides. Pectoral fins large, and upper tail-fin lobe much larger than lower lobe. Fin tips sometimes tipped blackish. **DISTRIBUTION** In Australian waters, ranges through south, from Moreton Bay Qld to around Geraldton WA. Worldwide in most subtropical and warm temperate oceans. **HABITAT AND HABITS** Inhabits depths from 2m to about 70m, including open seas, reefs, bays and estuaries, and also enters rivers. Active hunter of fish and turtles, but will also also investigate larger (potential) prey animals, and is attracted to splashing in the water caused by distressed or wounded animals. Although not usually aggressive, potentially dangerous to humans and has been responsible for at least one human fatality. Potentially Dangerous

Silky Shark ■ *Carcharhinus falciformis* TL 3.3m

DESCRIPTION Slender bodied with long, rounded snout, large pale eyes with dark pupils, and long, backwardly curved pectoral fins. Grey to bluish-grey above and white below. Dorsal fin located midway along back, behind line of pectoral fin-tips, and upper lobe of tail-fin heterocercal.
DISTRIBUTION In Australian waters, absent only from southern waters, from south-western WA, through tropical north to central NSW. Occurs in tropical and warm temperate marine waters worldwide. **HABITAT AND HABITS** Found on reefs and in open oceans, mainly at 200–500m depth, where it feeds on fish, squid and crabs. Aggressive, very fast moving and potentially dangerous to humans, although not commonly encountered by recreational water users. Potentially Dangerous

Galapagos Shark ■ *Carcharhinus galapagensis* TL 3.7m

DESCRIPTION Similar in appearance to the Grey Reef Shark (see p. 88), but longer and more slender, with a very erect dorsal fin. Pectoral fins large, and caudal fin strongly heterocercal. Brownish-grey to dark grey above and paler below. Darker fin-tips fade with

age. **DISTRIBUTION** Common around Lord Howe Island NSW and waters off WA. Occurs in tropical and temperate marine waters worldwide, and originally described from specimens caught around Galapagos Islands. **HABITAT AND HABITS** Inhabits shallow to deeper waters to around 285m depth, particularly around oceanic islands, and younger individuals are regularly seen in lagoons. Carnivorous, feeding mainly on fish or cephalopods taken mostly near the ocean floor. Very inquisitive towards humans, especially when they are spearfishing, and large individuals can be aggressive. Potentially Dangerous

Bull Shark ■ *Carcharhinus leucas* TL 3.4m

DESCRIPTION Stout-bodied, robust shark with a blunt, rounded snout and small eyes. Dark grey above and whitish below, with two dorsal fins, the second around one-third the height of the main one. **DISTRIBUTION** Northern Australia, from Swan River

WA to Clarence River NSW. Occurs in tropical and temperate waters worldwide. **HABITAT AND HABITS** Found in both marine and freshwater habitats, including coastal waters, river systems and lakes, often a long way from the coast. Hunts primarily for variety of aquatic prey, including other sharks, rays, turtles and echinoderms, but also birds and mammals, and will readily attack and kill humans. An aggressive shark with powerful jaws, it has been responsible for more than 30 human fatalities in the past 100 years, particularly in the murky waters of harbours and rivers. DANGEROUS

Common Blacktip Shark ■ *Carcharhinus limbatus* TL 2.75m

DESCRIPTION Dark bluish-grey to bronze above, with a broad darker stripe along sides, and cream to white below, with a distinct black spot on tips of pelvic fins. Most fins tipped with black, which fades with age. **DISTRIBUTION** Warm temperate, subtropical and tropical waters of Australia from Cape Leeuwin WA, north through NT and Qld, to Bermagui NSW. **HABITAT AND HABITS** Found in coastal and inshore waters to around 650m depth, including bays, estuaries and river outflows, generally near coral reefs. Feeds mainly on fish that are taken on or near the ocean floor, but will also take cephalopods and crustaceans. Not generally aggressive, and commonly seen by divers and snorkellers without incident, but must be regarded as potentially dangerous if agitated or when attracted to spearfishing activities. Potentially Dangerous

Oceanic Whitetip Shark ■ *Carcharhinus longimanus* TL 3m

DESCRIPTION Bronzed grey above and greyish-white below, with conspicuous white tips on most fins, including broad, rounded dorsal and pectoral fins. Younger individuals have darker fin-tips and black patch near tail. **DISTRIBUTION** Recorded off all Australian coasts, but less common in south and north. Occurs worldwide. **HABITAT AND HABITS** Found mainly in open tropical to warm temperate oceans at long distances from shore, to depths of around 150m, but can occur in inshore waters around oceanic islands. Hunts for fish, cephalopods, turtles and seabirds, and will feed on carrion. Although it is rarely encountered by humans and most interactions with divers are without incident, it is regarded as one of the most dangerous sharks and has been responsible for human fatalities. DANGEROUS

Blacktip Reef Shark ■ *Carcharhinus melanopterus* TL 1.4m

DESCRIPTION Yellowish-brown to grey above, with distinctive large black tips on first dorsal fin and lower caudal fins, and smaller black tips on most other fins. Dark longitudinal stripe along sides, behind first dorsal fin. **DISTRIBUTION** In Australia,

from around Shark Bay, WA, through northern Australia, to southern Qld. Found generally in tropical waters of Indo-Pacific. **HABITAT AND HABITS** Often associated with coral reefs that are mainly shallower than 40m where it feeds on fish, molluscs and crustaceans. Perhaps the most commonly encountered shark species in its preferred range and habitat, and not regarded as dangerous, but can be aggressive if threatened and must be treated with respect. Potentially Dangerous

Tiger Shark ■ *Galeocerdo cuvier* TL 6m

DESCRIPTION Mature individuals generally grey above, having lost the bold striped pattern of the young that gives the species its common name. First dorsal fin wide and angular, and second dorsal fin significantly smaller. **DISTRIBUTION** In Australia, occurs from Perth WA, through warmer northern waters, to southern NSW coast. Found in tropical waters worldwide, although not typically oceanic. **HABITAT AND HABITS**

Inhabits coastal waters, where it is largely nocturnal. Hunts variety of prey, including fish and turtles, and readily scavenges on larger mammals and other potential food items. Will also eat non-digestible items such as car-registration plates and tins. Regularly encountered in shallower waters, where it readily investigates humans, and has been identified in numerous fatal attacks. DANGEROUS

Whitetip Reef Shark ■ *Triaenodon obesus* TL 1.6m

DESCRIPTION Brownish-grey above, with conspicuous white tips to first dorsal fin and upper caudal fin. Second dorsal fin and lower caudal fin occasionally tipped with white.
DISTRIBUTION Northern Australia, from around Exmouth WA to southern Qld. Occurs in tropical marine waters throughout Indo-Pacific. **HABITAT AND HABITS** Found in coral reefs and rocky areas, where it shelters most of the day under ledges or in caves. Locates prey, mainly fish, at night using smell or electrical receptors, but will also feed opportunistically during the day and readily approach divers and spearfishers for handouts. An inquisitive species commonly seen by divers and snorkellers, with very few incidents. Usually considered harmless, but has been responsible for at least one unprovoked attack. Potentially Dangerous

Scalloped Hammerhead ■ *Sphyrna lewini* TL 4.3m

DESCRIPTION Brownish-grey to bronze above with darker tips to pectoral fins in adults, but otherwise unpatterned. Immatures have darker tips on pectoral, caudal and second dorsal fins. Head broad and mallet shaped. **DISTRIBUTION** In Australia, from north-western WA, through waters of NT and Qld, to Sydney NSW. Worldwide in tropical and warm temperate oceans. **HABITAT AND HABITS** Found in coastal and offshore waters, where it may occur in large schools, particularly when feeding in deep water on schools of fish and cephalopods. Although it is active at any time, most hunting takes place at night, with prey located using the electroreceptors and sensitive olfactory glands located along the broad head. Has been responsible for attacks on humans and should be regarded as potentially dangerous. Potentially Dangerous

Great Hammerhead ■ *Sphyrna mokarran* TL 4.5m

DESCRIPTION Grey-brown to greenish above and off-white below, with very tall first dorsal fin and broad, mallet-shaped head. Pectoral fins large and backswept, and most fins have dusky tips. **DISTRIBUTION** In Australia, from north-western WA, through waters of NT and Qld, to southern NSW. Worldwide in tropical and warm temperate oceans. **HABITAT AND HABITS** Found in coastal, inshore and oceanic marine waters, where

it can form large migratory shoals. Favours stingrays but also eats other fish species, including other species of shark, cephalopods and crustaceans. Most hunting takes place at night, with prey located using the electroreceptors and sensitive olfactory glands located along the broad head. Has been responsible for attacks on humans. Potentially Dangerous

Great Torpedo ■ *Tetronarce nobiliana* TL 1.05m

DESCRIPTION Brown to pinkish-brown above and with no discernable pattern. Underparts pale pinkish-white. Head has a broadly rounded disc and tail is short. **DISTRIBUTION** Throughout most offshore and oceanic waters of Australia, although seemingly absent from northern Qld and NT. Also off New Zealand and eastern Atlantic Ocean. **HABITAT AND HABITS** Inhabits sandy areas around reefs to around 800m depth, where it preys on fish and crustaceans at or near the sea bottom. Catches prey either by using an electric shock to stun it, or by concealing itself under the sand and seizing animals that come close enough. The electric shock is capable of causing severe muscle cramps in humans who handle or step on the fish, but it is generally not considered life-threatening. Potentially Dangerous

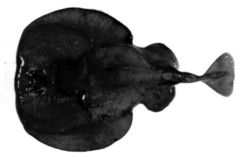

Ornate Numbfish ■ *Narcinops ornata* TL 24cm

DESCRIPTION Rounded head and long, muscular tail. Cream to pale brown above, distinctively patterned with variably sized darker pinkish-brown spots, each edged paler and occasionally merging together to form dark longitudinal bands. Underparts white. **DISTRIBUTION** Offshore waters of northern Australia from Gulf of Carpentaria Qld, west to northern WA. **HABITAT AND HABITS** Inhabits deeper oceanic waters up to depths of 50–130m, where

it feeds on small marine invertebrates along the ocean floor. Produces electric shocks as a defence against potential predators, and will shock humans if handled or provoked. Not considered fatal, but capable of causing severe muscle cramps. Potentially Dangerous

Tasmanian Numbfish ■ *Narcinops tasmaniensis* TL 47cm

DESCRIPTION Smooth, flattened body, with broadly blunted head, rounded at front and more angular at rear, and muscular, tapering tail. Brown above and white below, occasionally with dark markings. Young individuals often have a darker vertebral stripe. **DISTRIBUTION** Southern and eastern Australian waters, from northern NSW, to the Great Australian Bight WA and Tas. **HABITAT AND HABITS** Inhabits coastal offshore waters and the continental slope, up to about 600m depth in the north, but rarely exceeds depths of 100m around Tas. Produces a weak electric shock that it uses to stun its prey and deter potential predators. Not considered dangerous to humans, but can cause pain, discomfort and muscle cramps if handled. Harmful

Coffin Ray ■ *Hypnos monopterygius* TL 60cm

DESCRIPTION Colour varies in different environments, but generally pale brown to blackish. Flat bodied, with eyes and nostrils on dorsal surface and mouthparts on ventral

surface. Wings broad and rounded, merging with smaller, disc-shaped pelvic fins and short, broad tail. **DISTRIBUTION** Coastal and offshore waters from southern Qld to southern NSW, and from eastern SA to north-western WA. **HABITAT AND HABITS** Found in estuaries and deeper waters to around 240m depth, generally those with a soft sandy or muddy substrate. Conceals itself on the sea bottom leaving only its eyes and nostrils visible, and waits for fish to come within striking distance. Capable of powerful electric shocks, which are used to stun prey. Can survive for long periods if stranded by the tide, and should not be handled. Harmful

Brown Stingray ■ *Bathytoshia lata* TL 4m

DESCRIPTION Large stingray with sharp, thorn-like denticles along midline of upper surface. Flattened, with large pectoral fins and very long, 'whip-like' tail. Colour dark olive

to black above and whitish below. **DISTRIBUTION** Coastal and offshore waters of Australia, from Moreton Bay Qld, through NSW, Vic, Tas and SA, to around Port Hedland WA. **HABITAT AND HABITS** Inhabits subtropical to temperate marine waters, including sandy or muddy bays, harbours and reefs, where it feeds on crustaceans, molluscs, worms and fish along the ocean floor. Large venomous barb (occasionally two barbs) on tail-base is used solely for defence and can cause extreme pain and death in humans. Potentially Dangerous

Smooth Stingray ■ *Bathytoshia brevicaudata* TL 4.3m

DESCRIPTION Very large, flattened fish, with large pectoral fins (wings) forming a wide, somewhat ovate disc. Dark grey-brown above and whitish below with darker edges to wings. Head raised, with paler pores on either side. Body smooth, but tail long with numerous raised tubercles and a large, serrated barb. **DISTRIBUTION** Temperate coastal and offshore waters of southern Australia, from southern Qld, to around Shark Bay WA and southern Tas. **HABITAT AND HABITS** Bottom dwelling in inshore waters up to around 150m deep, where it feeds on variety of fish, crustaceans and molluscs. Often sighted in places where fish are cleaned. Large venomous barb can cause extreme pain and death in humans, with most envenomations occurring on the legs of swimmers who stand on the fish. Potentially Dangerous

Estuary Stingray ■ *Hemitrygon fluviorum* TL 1.3m

DESCRIPTION Sandy-brown above with broad, bluntly pointed wings, long, whip-like tail, and distinctive row of spines along back and base of tail. Head raised and snout broad and angular. **DISTRIBUTION** Along northern and eastern Australia from western NT, through Qld and south to around Merimbula NSW. **HABITAT AND HABITS** Inhabits inshore areas, including tidal estuaries, mangroves and swamps, to a depth of around 30m, where it hunts along the sea bottom for shellfish, worms and crustaceans. Large defensive venomous barb on upper side of tail can cause extreme pain in humans. Most envenomations occur on the feet or lower legs of swimmers who unwittingly stand on the fish. Potentially Dangerous

Ventral view, showing mouth and gills

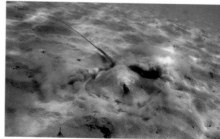

Foraging along the sea floor

Australian Whipray ▪ *Himantura australis* TL 3.5m

DESCRIPTION Pale brown, and distinctly spotted (young) or reticulated (subadults and adults) with darker brown or blackish-grey, and white below. Tail long and whip-like,

spotted at base and blackish from barb to tip, with faint paler marking on upper surface. **DISTRIBUTION** Northern Australian waters from around Brisbane Qld to around Shark Bay WA. **HABITAT AND HABITS** Mostly encountered in shallower waters, including bays and estuaries, up to a depth of about 45m. Large venomous barb on upper side of tail is used solely for defence against would-be aggressors and can cause extreme pain in humans. Most envenomations occur on the feet or lower legs of swimmers who unwittingly stand on the fish. Potentially Dangerous

Bluespotted Maskray ▪ *Neotrygon kuhlii* TL 70cm

DESCRIPTION Grey to grey-brown or greenish above with numerous blue-and-black spots, and whitish underparts. Tail long, with black cross-bands on terminal third and pale

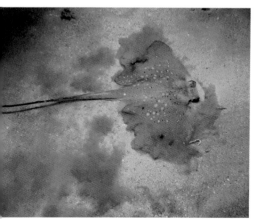

tip. Head raised, with large protruding eyes and dark face-mask. Tail broad down to single sting, but narrowing towards tip. **DISTRIBUTION** In Australia, from around Shark Bay WA, through NT and Qld, and south to northern NSW. Occurs throughout tropical marine waters of Indo-West Pacific. **HABITAT AND HABITS** Found in shallow coastal and inshore waters, normally at depths of less than 100m, but has been recorded at 170m. Bottom dwelling, often concealing itself beneath the sandy or muddy substrate, and feeds on crustaceans including crabs and shrimps. Venomous spine can inflict a painful wound. Potentially Dangerous

Bluespotted Ribbontail Ray ▪ *Taeniura lymma* TL 75cm

DESCRIPTION Olive to brownish or orange above, with numerous blue spots on oval-shaped disc and pelvic fins, and paler below. Tail moderately long, with two serrated barbs located towards tip, and edged with two blue longitudinal stripes. **DISTRIBUTION** Tropical marine waters from central WA, through NT and Qld, to northern NSW. More widely spread in Indo-West Pacific. **HABITAT AND HABITS** Common near coral reefs and seagrass beds in warm, shallow inshore waters generally less than 20m deep, often in large groups. Hunts in sandy areas on the incoming tides for worms and molluscs. Shelters at other times in caves and under rocky overhangs. Venomous spine can inflict a painful sting. Potentially Dangerous

Eastern Shovelnose Stingaree ▪ *Trygonoptera imitata* TL 80cm

DESCRIPTION Bottom-dwelling, flattened fish with large subcircular disc and tail slightly shorter than body length, with two venomous barbs. Yellowish-brown to greyish-brown above, darker along midline and paler towards edges of pectoral fins, with varying numbers of scattered darker spots. **DISTRIBUTION** South-eastern Australia, from Jervis Bay NSW, through Phillip Bay Vic and Bass Strait Tas, to St Vincent Gulf SA. **HABITAT AND HABITS** Inhabits bays, estuaries and offshore waters with sandy or muddy substrates, where it feeds on benthic invertebrates, including polychaete worms and small crustaceans, at depths of up to 120m. Venomous barbs are used for defence against potential predators, and are capable of causing intense pain, swelling and infection in humans. Venomous

Common Stingaree ■ *Trygonoptera testacea* TL 47cm

DESCRIPTION Greyish-brown to dark brown above and whitish below, with dorsally flattened body, raised head section with eyes located on either side, and small dorsal fin.

Tail has one or two venomous spines. **DISTRIBUTION** Along eastern Australia, from southern Qld to southern NSW. Very common in east (less common in south). **HABITAT AND HABITS** Found in shallow coastal and deeper inshore waters to about 135m depth, where it hunts along the ocean floor around reefs for small crustaceans and marine worms. Perhaps the most commonly encountered species in most of its range. Care should be taken when handling it as venomous spines on the posterior portion of the tail are capable of causing a very painful injury in humans. Venomous

Banded Stingaree ■ *Urolophus cruciatus* TL 50cm

DESCRIPTION Readily identified by yellowish-brown colour with variable contrasting blackish cross-bars and patches on body, extending on to wings, and longitudinal blackish stripe on top of short tail.

DISTRIBUTION Temperate waters of south-eastern Australia, from Jervis Bay NSW, along Vic and Tas, to around Robe SA. **HABITAT AND HABITS** Found in shallow and deeper waters around bays and estuaries with sandy or muddy substrates, at 1–200m depth. Generally partially buried in the sand during the day, and emerges at night to hunt for small crustaceans and benthic worms. The single, serrated spine towards the tail-tip is used for defence, and is capable of injecting a potent venom. If handled, great care should be taken as the venom can cause extreme pain in humans. Venomous

Kapala Stingaree ▪ *Urolophus kapalensis* TL 45cm

DESCRIPTION Mostly unpatterned greenish to pinkish-brown, with conspicuous darker mask over eyes and near pelvic fins. Tail has small dorsal fin and serrated spine towards tip.
DISTRIBUTION Restricted to waters of eastern Australia, from Moreton Bay Qld to southern NSW. **HABITAT AND HABITS** Found in coastal and inshore areas to depths of about 85m, where it forages around reefs and seagrass meadows for prey, presumably small crustaceans and polychaete worms. Has a serrated venomous spine towards the tail-tip, which is used for defence and is capable of causing extreme pain in humans. Venomous

Whitespotted Eagle Ray ▪ *Aetobatus ocellatus* TL 1.53m

DESCRIPTION Dark greenish-grey above, occasionally tinged pink, with numerous greyish-white to pale blue spots and circles. Tail very long and thin, with long barbs on upper side at its base.
DISTRIBUTION Ranges from around Sydney NSW, through tropical northern Australian waters, to Shark Bay WA. **HABITAT AND HABITS** Found in warmer coastal waters, including bays and estuaries, and deeper open oceans, where it can congregate in large schools. Often associated with coral reefs, where it actively hunts for cephalopods, crustaceans, molluscs, fish and marine worms. Venomous tail spines are used solely for defence against potential predators, and are capable of causing significant pain and injury to humans. Venomous

Southern Eagle Ray ▪ *Myliobatis tenuicaudatus* TL 2.4m

DESCRIPTION Variably coloured with grey, olive-green, brownish or yellowish above, and patterned with wavy greyish-blue barring and blotches. Tail long and whip-like, with

small dorsal fin at base and venomous spine. **DISTRIBUTION** Coastal and offshore waters of southern Australia, from southern Qld to around Shark Bay WA, and to waters off southern Tas. Range also extends to NZ. **HABITAT AND HABITS** Found in bays and sheltered coastal seagrass meadows, where it actively hunts for molluscs, crustaceans and marine worms, which are caught on or near the ocean floor. Prey can also be found by digging in the substrate. Venomous barb is used for defence and can cause extreme pain in humans. Venomous

Green Moray ▪ *Gymnothorax prasinus* TL 1m

DESCRIPTION Yellow to olive-brown with orange head, and large mouth containing several large, pointed teeth. Individuals can appear green in shallow waters due to

green algal growths on the skin. **DISTRIBUTION** In Australia, from southern Qld, south along NSW, Vic, northern Tas and parts of SA, to around Shark Bay WA. Occurs in tropical and temperate oceans worldwide. **HABITAT AND HABITS** Found in estuaries and along rocky reefs at 1–30m depth, where it inhabits hollows and crevices that are occasionally shared by two or more individuals. Generally shy in shallow pools, but more inquisitive in deeper water and will emerge to investigate divers. Capable of inflicting a painful bite if threatened, and if the wound is not treated properly it can become infected. Harmful

Undulate Moray ■ *Gymnothorax undulatus* TL 1m

DESCRIPTION Boldly patterned with large blotches in browns and blacks along body, and yellowish head with long snout and sharp, slender teeth. **DISTRIBUTION** Temperate coastal waters of Indo-Pacific, from around Houtman Albrolhos WA to Cape Byron NSW. **HABITAT AND HABITS** Found on rocky reef slopes to about 25m depth, where it is largely nocturnal. Actively hunts a variety of fish species, octopuses and crustaceans that are located largely by smell as well as by investigating burrows and crevices. Known for aggressiveness towards divers, and capable of inflicting a painful bite. Consumption of the flesh and liver can lead to severe ciguatera poisoning, which can result in death if a sufficient quantity of the toxin is ingested. Harmful

Estuary Catfish ■ *Cnidoglanis macrocephalus* TL 60cm

DESCRIPTION Generally brown, heavily mottled with yellow and black, with flattened head, large lips and outwards protruding nostrils. Four pairs of barbels around mouth. Body elongated, with long, tapering tail with continuous dorsal, anal and caudal fins. **DISTRIBUTION** Temperate marine waters of southern Australia, from southern Qld, through Vic and SA, to around Geraldton WA. **HABITAT AND HABITS** Found in estuaries and coastal waters with muddy substrates, down to around 40m depth. Mainly nocturnal, sheltering during the day in crevices and under ledges, and foraging at night for crustaceans, molluscs and worms. Single serrated venomous spine in each of the dorsal and pectoral fins can inflict a painful wound in humans. Venomous

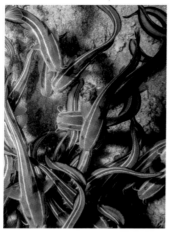

Striped Catfish
■ *Plotosus lineatus* TL 35cm

DESCRIPTION Greyish to dark brown with two pairs of paler longitudinal stripes, which are more visible in younger individuals than in adults. Concealed venomous dorsal and pectoral spines. **DISTRIBUTION** In Australia, in tropical and subtropical waters from south-western WA, around NT and Qld, to Sydney NSW. Widespread in Indo-West Pacific region and Mediterranean Sea. **HABITAT AND HABITS** Found in seagrass beds, rocky reefs, and open sandy or muddy areas in bays, estuaries and intertidal pools to depths of around 30m (occasionally to 75m). Can occur in large numbers, and feeds on invertebrates, polychaete worms, algae and some fish. Venom causes intense pain in humans, which can last for several hours. Venomous

Eastern Frogfish ■ *Batrachomoeus dubius* TL 30cm

DESCRIPTION Mottled with pale grey to bluish-grey and brown. Large eyes and spines on dorsal surface and gill coverings. Head wide and flattened, with large mouth and fleshy lips. Juveniles pale with dark bands. **DISTRIBUTION** Endemic to east coast of Australia, from central NSW to southern Qld. **HABITAT AND HABITS** Inhabits rocky reefs in deeper parts (down to 150m) of bays and harbours, where it conceals itself under the silty or sandy floor, or within rock holes, and waits for crustaceans and fish to come within range. Its large mouth and expandable stomach enable it to swallow large prey. The three short spines of the dorsal fin and twin spines on the gill coverings are considered venomous. Venomous

Little Gurnard Perch ■ *Maxillicosta scabriceps* TL 12cm

DESCRIPTION Mottled yellowish-brown to reddish-brown, grey and black above, with black patch in dorsal fin, between fifth and ninth spines, and large eyes. Underparts whitish. **DISTRIBUTION** Coastal and inshore waters of southern Australia, from central Vic to central WA. **HABITAT AND HABITS** Bottom-dwelling fish of inshore waters with sandy or stony substrates, where it favours seagrass meadows, occurring at depths of 2–46m. Nocturnal, spending most of the time buried under the sand during the day. Venomous spines can inflict a painful wound in humans. Venomous

Common Gurnard Perch ■ *Neosebastes scorpaenoides* TL 35cm

DESCRIPTION Well camouflaged with its surroundings, with mottled pattern of browns, pinkish-red and blackish. Terminal part of caudal fin and undersurface of pectoral fins are white. Head proportionally larger in adults than in juveniles, giving a snub-nosed appearance. **DISTRIBUTION** Waters off south-eastern Australia, from southern NSW, through Vic and Tas, to Great Australian Bight SA. **HABITAT AND HABITS** Found in coastal waters with sandy substrates and reefs down to around 140m depth. Prefers sandy patches among sponge gardens adjacent to rocky reefs. Crepuscular and nocturnal, hunting for small crustaceans, cephalopods and fish. Venomous spines can cause severe pain in humans. Venomous

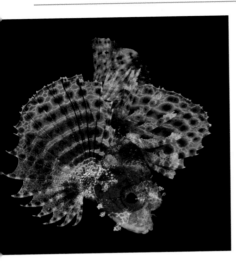

Zebra Lionfish
■ *Dendrochirus zebra* TL 25cm

DESCRIPTION White to pinkish or reddish with five or six broad darker vertical reddish-brown bars on sides of body, and dark spot on cheek. Large individuals have narrower bands alternating with broad bands. Long dorsal and pectoral fins. **DISTRIBUTION** In Australia, from Shark Bay WA, through waters of NT and Qld, to around Jervis Bay NSW. More widely distributed throughout Indo-Pacific. **HABITAT AND HABITS** Found in marine waters with rocky substrates and around rocky reefs, normally down to 60m depth. Shelters during the day under rocky ledges and in crevices and caves. Long dorsal spines are venomous. Venomous

Common Lionfish ■ *Pterois volitans* TL 35cm

DESCRIPTION Colour variable, generally pale with numerous reddish to black vertical bands. Pectoral and dorsal fins very long. Long tentacle above eyes, usually longer in young fish than in adults and becoming more leaf-like with age, but may be absent in some adults.

DISTRIBUTION In Australia, from south-western WA, around NT and Qld, to south of Sydney NSW. Occurs in tropical waters throughout Indo-Pacific. **HABITAT AND HABITS** Inhabits tropical reefs in coastal bays and offshore waters. Possesses 13 dorsal, three anal and one pelvic venomous spines, all capable of causing extreme and immediate pain, but is not aggressive and is often approached by divers for photography. Venomous

Eastern Red Scorpionfish ▪ *Scorpaena jacksoniensis* TL 40cm

DESCRIPTION Colour highly variable and it is well camouflaged in its surroundings, but usually bright red with bands and large patches of darker reds, greys and whites. Often has loose, leafy skin-flaps on body. Nine dorsal fin rays and 12 dorsal spines. **DISTRIBUTION** Coastal and offshore temperate waters of eastern Australia, from southern Qld to eastern Vic. **HABITAT AND HABITS** Inhabits rocky, sponge-covered reefs up to around 40m deep, and coastal estuaries. Feeds on fish and invertebrates that are normally swallowed whole. Lies motionless on the bottom, waiting for prey to come within range. Popular fish commonly caught by anglers fishing over shallow reefs, and easily approached by divers. Venomous spines can cause localized pain that can spread through an entire limb and into the chest, and occasionally cause vomiting or fainting. Venomous

Southern Red Scorpionfish ▪ *Scorpaena papillosa* TL 20cm

DESCRIPTION Colour highly variable, with mottled pattern of dark brown, reddish-brown, black and white, and darker rear edges to scales of body. Ten to 11 dorsal fin rays. **DISTRIBUTION** Waters of southern Australia, from northern NSW, through Vic, Tas, SA and south-western WA. Also in NZ. **HABITAT AND HABITS** Inhabits coastal and offshore waters mainly at 10–130m depth, where it feeds on small fish and shellfish. Humans who come into contact with the venomous spines suffer from localized pain that can spread through an entire limb and into the chest. Some victims also suffer from vomiting or fainting. This species may comprise up to four distinct species, but is treated as a single species here pending further classification. Venomous

Raggy Scorpionfish ▪ *Scorpaenopsis venosa* TL 25cm

DESCRIPTION Cryptically coloured and patterned to aid camouflage, which is further enhanced by the numerous skin-flaps and tentacles. Adults have scattered, small pale blue

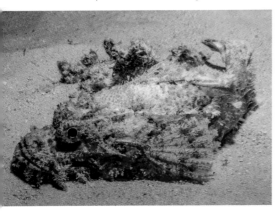

dots on body and dark triangle under each eye. **DISTRIBUTION** In Australia, from islands of north-western WA, through NT to southern Qld. Occurs generally in Eastern Indian and Western Pacific Oceans. **HABITAT AND HABITS** Found in sandy and muddy areas around coastal reefs, usually concealed within algae, sponge gardens and seaweed. Bottom dwelling, most commonly at depths of 2–25m, and has been recorded down to 95m, but rarely far away from coasts and inshore islands. Venomous spines can inflict a painful sting. Venomous

Leaf Scorpionfish ▪ *Taenianotus triacanthus* TL 10cm

DESCRIPTION Body strongly laterally compressed, with large, erectile dorsal crest and leaf-shaped tail. Main colour variable, including white, pale brown, yellow, green, pink

and reddish, often with darker and paler spots, streaks and blotches. **DISTRIBUTION** Tropical waters of Indo-Pacific, including northern Australia, from Pilbara coast WA to northern NSW coast. **HABITAT AND HABITS** Inhabits sandy and muddy areas on shallow coastal reefs and occasionally deeper waters, where it conceals itself among seagrass and algae. Resembles a fallen leaf, and sways from side to side. The 12 dorsal and three anal spines are venomous and can inflict an extremely painful wound that can persist for several days, and be accompanied by sweating, headaches and potentially a severe allergic reaction. Venomous

Eastern Fortescue ■ *Centropogon australis* TL 15cm

DESCRIPTION Highly variable, being able to change colouration for camouflage. Commonly pale grey with incomplete broad vertical bands or elongated blotches of dark brown to blackish. **DISTRIBUTION** Waters of eastern Australia, from central Qld to southern NSW. **HABITAT AND HABITS** Often encountered in estuaries, particularly when breeding, but also on inshore reefs. Forages alone or in groups on sandy, rocky or coral substrates, and in sponge gardens. Spines capable of inflicting a very painful sting on humans, but stings rarely occur underwater. Pain from a sting can be severe, lasting up to an hour or more, with most interactions occurring when the fish is an unwanted bycatch of fishing or prawning activities. Venomous

Goblinfish ■ *Glyptauchen panduratus* TL 20cm

DESCRIPTION Variably coloured, including white, brown and black, and able to change colour to camouflage itself in its environment. Pale patches on body and fins have black spotting. Has 17 tightly spaced spines enclosed within its dorsal web. **DISTRIBUTION** Southern Australia, from Perth WA, through southern states, to northern NSW. **HABITAT AND HABITS** Found in coastal and inshore marine waters and estuaries down to depths of around 60m, where it inhabits rocky areas. Mainly nocturnal, remaining hidden during the day and emerging at night to hunt. Venom is injected through contact with the dorsal spines and can cause intense pain, shock and loss of consciousness. Venomous

Soldier ■ *Gymnapistes marmoratus* TL 22cm

DESCRIPTION Pale purplish-white to yellowish-brown with mottling of darker brown and purplish-brown. Thirteen venomous dorsal spines, two large backwards facing ones near

mouth, one on chest and three towards underside of tail. Body does not have scales. **DISTRIBUTION** Temperate waters of southern Australia, from central NSW, through Vic, Tas and SA, to around Geraldton WA. **HABITAT AND HABITS** Found at variable depths of 2–20m, where it inhabits rocky reefs, seagrass beds and other environments on the ocean floor, but is most commonly encountered in inshore water estuaries. Large numbers form in shallow water before breeding. Venomous spines are capable of inflicting a painful sting if a fish is handled or encountered while wading through water. Venomous

Bullrout ■ *Notesthes robusta* TL 20cm

DESCRIPTION Mottled browns, yellows and olives. Large, scaleless head, bony ridge above eye, and operculum (gill covering) with seven spines. **DISTRIBUTION** Coastal

waters of eastern Australia from Annan River Qld, south to Clyde River NSW. **HABITAT AND HABITS** Inhabits bays and estuaries, and commonly encountered in lower tidal reaches of rivers and creeks, where it shelters among vegetation and around submerged tree roots or rocky areas. Ambush predator of crustaceans and small fish. Tends not to swim away when disturbed, instead raising its venomous spines, which cause severe pain in humans within seconds of envenomation. The pain can become excruciating, with localized swelling occasionally occurring around the wound site and duller pain in the lymph-node regions. Venomous

Reef Ocean Perch ▪ *Helicolenus percoides* TL 25cm

DESCRIPTION Orange and white, with two darker vertical bands on sides that both pair off towards upper limit. Further two bands on tail. Dorsal fin ray deeply notched and pectoral fins large. **DISTRIBUTION** Waters of southern Australia, from northern NSW, Vic, Tas and SA, to around Geraldton WA. Also in NZ. **HABITAT AND HABITS** Inhabits deeper marine waters mainly at 80–150m depth, though it can reach 300m, where it lies concealed within seagrass. Seizes shrimp, squid and smaller fish that come within range overhead. Single female can produce more than 150,000 larvae in a season. Venomous spines can inflict a painful injury in humans. Venomous

Demon Stingerfish ▪ *Inimicus caledonicus* TL 25cm

DESCRIPTION Variable colours and patterns, and well camouflaged in its surroundings, but normally brown, occasionally with paler blotches. Pectoral fins blackish, with broad, curved pale stripe and numerous yellow flecks and spots. **DISTRIBUTION** In Australia, along eastern Qld and far north of NSW. Generally in Indian and Pacific Oceans around Australia, New Caledonia, PNG, Thailand and India. **HABITAT AND HABITS** Found in shallow tropical marine waters at 15–60m depth, where it conceals itself among rocky, sandy or muddy substrates on the ocean floor, usually around coral reefs. Ambush predator of smaller fish and crustaceans, and can crawl along the ocean floor using its modified pectoral fins. Has extremely venomous spines, which are capable of causing intense pain and potential death in humans. DANGEROUSLY VENOMOUS

Estuarine Stonefish ■ *Synanceia horrida* TL 35cm

DESCRIPTION Generally brownish-grey to reddish-brown, with yellow to red patches sometimes visible, and excellently camouflaged, appearing to be nothing more than a lump

of coral or rock to a casual observer. Eyes elevated and separated by bony ridge. **DISTRIBUTION** In Australian waters, from far northern NSW, through Qld and NT, to Shark Bay WA. Occurs generally in tropical waters of Indo-Pacific. **HABITAT AND HABITS** Bottom-dwelling inhabitant of marine reefs at 0–40m depth, where it is an ambush predator of fish and crustaceans. If disturbed it seldoms swims away, instead raising its 13–14 dorsal spines for protection and injecting a highly toxic venom. One of the most venomous fish in the world and has been responsible for many deaths, although no deaths have been recorded in Australia. DANGEROUSLY VENOMOUS

Reef Stonefish ■ *Synanceia verrucosa* TL 35cm

DESCRIPTION Generally brown to grey, with yellow to red patches sometimes visible, and excellently camouflaged as a lump of coral or rock. Elevated eyes separated by a

depression. **DISTRIBUTION** In Australian waters, from far northern NSW to northern Qld, northern NT and much of WA. Occurs generally in tropical waters of Indo-Pacific. **HABITAT AND HABITS** Found in marine reefs, where it conceals itself among coral or on rocky ocean floor and waits for fish and crustaceans to come within range. If disturbed seldoms swims away, instead raising its 13 venomous dorsal spines and injecting a highly toxic venom into the intruder. One of the world's most venomous fish, responsible for many deaths, although no deaths have been recorded in Australia. DANGEROUSLY VENOMOUS

Southern Sand Flathead ■ *Platycephalus bassensis* TL 46cm

DESCRIPTION Yellowish-brown above, with numerous darker spots and obscure dark cross-bands, and white below. Head greatly flattened, with two rear-facing spines between cheek and gill covering, or preoperculum. Caudal fin has distinctive dark blotch, bordered by pale brown or white, on lower half. **DISTRIBUTION** Temperate marine waters of coastal southern Australia, from central NSW, through Vic, Tas and SA, to southern WA. **HABITAT AND HABITS** Found in sandy bays and estuaries to depths of around 100m, where it forms small, loosely formed groups. Buries itself beneath sand, leaving only the eyes visible, and waiting for prey, mainly small fish, squid and crustaceans, to come within striking distance. Spines are mildly venomous to humans. Venomous

Dusky Flathead ■ *Platycephalus fuscus* TL 1.2m

DESCRIPTION Largest Australian flathead species, with variable colouration of pale yellowish-brown with darker spots, to dusky brown. Pectoral fins have rows of small dark spots, and lower part of caudal fin is bluish. Head greatly flattened, with two rear-facing spines on preoperculum. **DISTRIBUTION** Coastal waters of eastern Australia, from southern Qld to eastern Vic. **HABITAT AND HABITS** Inhabits shallow sandy or muddy bays and rocky reefs to 25m depth, where it buries itself beneath sand, leaving only its eyes visible. Ambush predator of small fish, squid and crustaceans, lying in wait until prey comes within striking distance and swallowing it whole. Popular fish with anglers, and care must be taken to avoid the venomous spines when handling it. Venomous

Old Wife ■ *Enoplosus armatus* TL 25cm
(Bastard Dory)

DESCRIPTION Adults predominantly silver with 6–8 bold black, zebra-like stripes, and two elongated sections of dorsal fin. Young blotched darker, and have a white-edged spot on a single dorsal fin.

DISTRIBUTION Common in coastal and deeper waters of southern Australia, from southern Qld in east to around Geraldton WA in west. **HABITAT AND HABITS** Found in still or slow-moving waters of bays and protected areas of offshore reefs, where it feeds individually, in pairs or in shoals on plankton and small crustaceans. Young tend to occupy patches of weed on reefs or in estuaries. Dorsal fins of adults have venomous spines that can inflict a painful sting if the fish is handled incorrectly. Venomous

Great Barracuda ■ *Sphyraena barracuda* TL 1.4m

DESCRIPTION Silvery-grey with darker diagonal bars (upper) and blotches (lower) on sides. Caudal fin is scalloped and dorsal fin is triangular, and both have pale tips. Head elongated, with a large mouth and numerous sabre-like teeth. **DISTRIBUTION**

In Australia, from south-western WA, through tropical north, to southern NSW. Widespread in Indian and Pacific Oceans. **HABITAT AND HABITS** Found in open marine waters and around coral reefs, although younger fish stay closer inshore. Skilled and ruthless marine predator of fish, cephalopods and prawns. Consumption of larger individuals can cause ciguatera poisoning in humans, and the large, razor-sharp teeth can inflict nasty wounds if the fish is handled. Harmful

Common Stargazer ▪ *Kathetostoma laeve* TL 75cm

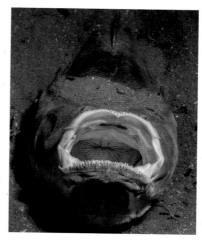

DESCRIPTION Normally grey-brown above, often with two darker cross-bands behind head, and white below. Pectoral and caudal fins have white margins. Head large and square, with large, upwards facing mouth. **DISTRIBUTION** Temperate marine waters of southern Australia, from northern NSW, through Vic, Tas and SA, to around Albany WA. **HABITAT AND HABITS** Inhabits shallow coastal bays and estuaries usually to 60m depth, but can reach 150m, where it lies concealed beneath the substrate with just its mouth and eyes exposed. Ambush predator of fish and crustaceans. Large, rear-pointing defensive spine on each operculum is venomous, but does not normally inflict any damage unless the fish is handled. Can be aggressive towards divers if disturbed at night. Venomous

Linespot Fangblenny ▪ *Meiacanthus grammistes* TL 11cm

DESCRIPTION White, with longitudinal black-and-yellow lines from head to just short of tail. Dorsal crest has black marking on outer edge of fin ray and caudal fins have rows of black spots. **DISTRIBUTION**

In Australia, from north-western WA to GBR Qld. Found generally in Indian and Pacific Oceans. **HABITAT AND HABITS** Occurs in sheltered reefs and lagoons, where it is largely solitary, actively foraging for plankton, fish eggs, algae and smaller fish. The genus is unique among fish in having specialized toxic buccal glands – the grooved canines deliver the toxin, although it appears to be used solely for defence. Venom is not thought to cause pain in humans, but can result in inflammation and dizziness. Harmful

Eyestripe Surgeonfish ■ *Acanthurus dussumieri* TL 50cm

DESCRIPTION Oval-shaped, mustard yellow-and-blue fish, with numerous wavy blue lines or spots mainly on head, large yellow patch behind eye and yellow band across

forehead. **DISTRIBUTION** In Australia, from around Exmouth WA, through NT and Qld, to central NSW coast. Occurs generally in tropical marine waters of Indo-Pacific. **HABITAT AND HABITS** Found in wide range of habitats, but favours reefs at depths of down to 130m, where it feeds on sediments and detritus. Sharp tail-spine, used in combat or to protect the tail from predatory fish, is capable of inflicting large, painful wounds in humans, and anglers should use a glove or towel when handling this fish if it is caught. Harmful

Striped Surgeonfish ■ *Acanthurus lineatus* TL 38cm

DESCRIPTION Yellow to orange on uppersides, with blue longitudinal stripes edged with black, and pale blue to purple on belly. Tail strongly crescent shaped (less so in younger individuals). **DISTRIBUTION** In Australian waters, from north-western WA, through

NT and Qld, to northern NSW. Widespread in tropical and subtropical waters throughout Indo-Pacific. **HABITAT AND HABITS** Occurs in shoals in shallow waters around exposed reefs, where it shelters at night in gullies and crevices, emerging to feed in guarded territories, mainly on red and green algae. Sharp tail-spine, used in combat or to protect tail from predatory fish, is capable of inflicting large, painful wounds in humans, and anglers should use a glove or towel when handling this fish if it is caught. Harmful

Goldlined Rabbitfish ■ *Siganus lineatus* TL 45cm

DESCRIPTION Colour bright blue on back and sides, and silver on belly, with bright yellow spot towards tail. Fifteen longitudinal golden-yellow lines along body. **DISTRIBUTION** Tropical marine waters of northern Australia, from Port Hedland WA, around NT, to far southern Qld. **HABITAT AND HABITS** Occurs on shallow coastal reefs down to 25m depth, although young are more prevalent in estuaries, around seagrass flats and mangroves. Feeds in schools of varying sizes on algae and sponges. When spawning, schools can grow to several thousand individuals, but typical shoals are much smaller – they become smaller, with up to around 25 individuals, as the fish mature. The 13 dorsal, seven anal and four pelvic spines contain a venom that can cause intense pain in humans. Venomous

Scribbled Rabbitfish ■ *Siganus spinus* TL 28cm

DESCRIPTION Colour variable, including creamish, greyish-blue, brown and blackish, with network of wavy lines on upper body, becoming more longitudinal on lower body. Eyes golden-yellow with brown cross. **DISTRIBUTION** In Australian waters, tropical north, from Shark Bay WA to Gladstone Qld. Occurs generally in marine waters of Indo-West Pacific. **HABITAT AND HABITS** Found around reefs to a depth of around 50m, but more commonly at 1–20m, where it is usually seen feeding on algae in small to medium-sized shoals. When threatened, lowers head to expose first dorsal spine. The 13 dorsal, seven anal and four pelvic spines contain a venom that can cause intense pain in humans. Venomous

Titan Triggerfish ■ *Balistoides viridescens* TL 75cm

DESCRIPTION Large, stocky fish with coarse scales and a deep groove forwards of eye. Body scales have dark centres, forming a mosaic pattern, and dorsal, anal and caudal fins are yellowish with blackish margins. Head elongated, with broad patches of yellow and white, and yellow-spotted, black patch between eye and pectoral fin. **DISTRIBUTION**

In Australia, in waters off north-western WA, through NT and Qld, to central NSW coast. Widespread in Indo-Pacific. **HABITAT AND HABITS** Found in shallow tropical waters, mainly around coral reefs, where it feeds on sea urchins, coral, molluscs, crustaceans and polychaete worms. Aggressive towards intruders when guarding nest, and will readily attack and bite divers who stray too close. Harmful

Thornback Cowfish ■ *Lactoria fornasini* TL 45cm

DESCRIPTION Yellowish-green to brown with numerous blue lines and spots. Paler below. Body box shaped and bony, with two large spines projecting forwards above eyes, single dorsal spine and pair of spines facing backwards near caudal fins. **DISTRIBUTION** In Australia, known from east and west coasts, from north-western to south-western WA,

and from north-eastern Qld to eastern Vic. Generally in tropical (mainly) and temperate marine waters of Indo-Pacific. **HABITAT AND HABITS** Found on coral and rocky reefs, where it is largely solitary, feeding on small molluscs, polychaete worms, small fish and algae. Secretes a neurotoxin from its skin when stressed, so care should be taken if handling it, even when dead. Poisonous

Yellow Boxfish ■ *Ostracion cubicus* TL 45cm

DESCRIPTION Yellow with brown to black spots, which are larger and darker on younger fish, and smaller and more brownish (occasionally whitish) in adults. Body box shaped and bony. **DISTRIBUTION** In Australia, waters off central Wa, through NT and Qld, to south of Sydney NSW. Generally in tropical and temperate waters of Indo-West Pacific. **HABITAT AND HABITS** Inhabits shallow coral and rocky reefs to depths of around 40m, where it feeds on small molluscs, crustaceans, polychaete worms and algae. Secretes a neurotoxin from its skin when stressed, so care should be taken if handling it. Although the toxin can only be secreted when the fish is alive, dead fish may also have residues of it on the body. Poisonous

Smooth Toadfish ■ *Tetractenos glaber* TL 15cm

DESCRIPTION Yellowish-brown above with numerous blackish spots and blotches on back and upper sides, often pooling together to form bands. Underparts white and tail reddish. Skin mostly smooth, with only very small spines. **DISTRIBUTION** South-eastern Australia, from Moreton Bay Qld, west to Kangaroo Island SA, and south to southern Tas.

HABITAT AND HABITS Occurs in coastal bays and estuaries, occasionally entering freshwater rivers to around 3m. Feeds close to the bottom on invertebrates and plankton, burying itself in sand at other times and leaving only its eyes exposed, and inflating its body with water when threatened. Several fatalities have been caused by consumption of this species due to the tetrodotoxin contained in its flesh and organs. Poisonous

Common Toadfish ■ *Tetractenos hamiltoni* TL 13cm

DESCRIPTION Mainly pale brown to yellowish-brown above, with numerous variably sized darker brown spots on dorsal surface and larger blotches on sides. Lower sides and

underparts whitish. **DISTRIBUTION** Resident on east coast of Australia, from Townsville Qld to Merimbula NSW. **HABITAT AND HABITS** Occurs in shallow coastal bays and estuaries to around 3m depth, where it is commonly sighted, particularly around jetties. Like other pufferfish species, inflates body with water when threatened. Several fatalities have been caused by consumption of this species due to the tetrodotoxin contained in its flesh and organs. Common pest to anglers that is regularly caught and released safely, but people are advised to wash their hands thoroughly after contact. Poisonous

Threebar Porcupinefish ■ *Dicotylichthys punctulatus* TL 43 cm

DESCRIPTION Greenish to bluish above, with dark brownish markings on dorsal surface,

below eye, through ear and midway along sides, forming three distinct vertical bars. White below. Numerous small black spots and many spines over body. **DISTRIBUTION** Temperate waters along east coast of Australia, from southern Qld to eastern Vic, and islands of north-eastern Tas. **HABITAT AND HABITS** Found in coastal and offshore waters, usually near rocky reefs, where it feeds on small marine invertebrates. Flesh contains the toxin tetrodotoxin, and can cause fatalities in humans if eaten. Poisonous

Freckled Porcupinefish ■ *Diodon holocanthus* TL 35 cm

DESCRIPTION Pale pinkish-brown above, with numerous small dark 'freckles', and larger patches of light brown above large eyes and along dorsal surface. Creamish below. Body covered in long, erectable spines.
DISTRIBUTION. In Australia, from south-western WA, through NT and Qld, to southern NSW. Occurs in marine waters worldwide.
HABITAT AND HABITS Found in coastal and estuarine waters with muddy or sandy substrates, in vicinity of rocky reefs, generally down to 35m depth, where it feeds at night on molluscs, crustaceans and sea urchins. If threatened, inflates body with water to form a large ball shape with spines fully erected. Flesh contains the toxin tetrodotoxin, and can be fatal if eaten. Poisonous

Slender-spined Porcupinefish ■ *Diodon nicthemerus* TL 30cm

DESCRIPTION Brownish above, with 3–4 darker vertical bars on sides, and white below. Body covered with numerous long white or yellowish spines, which are normally held flat against body but are extended when alarmed. Eye large with yellow ring. **DISTRIBUTION** Temperate waters from northern NSW, through Vic, Tas and SA, to south-western WA. **HABITAT AND HABITS** Found in shallow bays to a maximum depth of 85m, where it feeds nocturnally on crustaceans. When threatened, inflates body with water to form a large ball shape with spines fully erected, making it difficult to grasp by its attacker. Flesh contains the toxin tetrodotoxin, and can be fatal if eaten. Poisonous

Cane Toad ▪ *Rhinella marina* TL 12cm

DESCRIPTION Adults yellow to brown or greyish above with black spotting. Juveniles grey to black, often flecked with orange and brown. Underparts paler and marbled. **DISTRIBUTION** Constantly changing and expanding. Currently south to north coast NSW and west to Kimberleys WA. **HABITAT AND HABITS** Found in various habitats including mangroves, swamps, forests, grassland, rocky gorges, and agricultural and urban

areas. Mostly nocturnal as adults, feeding mainly on insects and spiders, but also small vertebrates. Female lays up to 35,000 black eggs in long strands. Poison is produced in glands all over body (more concentrated behind eyes). If ingested it can kill large animals, and cause severe illness in humans. Most injuries to people occur when a toad is squeezed, causing toxin to squirt into their eyes. Poisonous

Southern Corroboree Frog ▪ *Pseudophryne corroboree* TL 3cm

DESCRIPTION Top of body striped with black and yellow. Ventral surface black and white. **DISTRIBUTION** Currently restricted to single site in Mt Kosciusko, but formerly found throughout Australian Alps. **HABITAT AND HABITS** Occurs in sphagnum bogs, where it mainly feeds on small arthropods such as ants. Females lay their eggs above the

water and guard them until water levels increase in the depression, inundating the clutch and causing the eggs to hatch. The frogs produce alkaloid toxins that are similar to those of the poison dart frogs of South America; they also sequester alkaloid toxins from their food. The toxin is thought to be harmful to mammals. Poisonous

Southern Toadlet ■ *Pseudophryne semimarmorata* TL 3cm

DESCRIPTION Top of body grey to blue and occasionally metallic green with darker spots and blotches. Body covered in small tubercles. Ventral surface orange or yellow with black and white. **DISTRIBUTION** Eastern Tas and southern Vic. **HABITAT AND HABITS** Found in grassland, heaths and forests in areas that form waterlogged depressions, where it mainly feeds on small arthropods such as ants. Females lay their eggs above the water and guard them until water levels increase in the depression, inundating the clutch and causing the eggs to hatch. The frogs produce alkaloid toxins that are similar to those of the poison dart frogs of South America; they also sequester alkaloid toxins from their food. The toxin is thought to be harmful to mammals. Poisonous

Golden-tailed Gecko ■ *Strophurus taenicauda* TL 14cm

DESCRIPTION White or grey with black reticulations encompassing body. Tail has a ragged-edged yellow, orange or red stripe that sometimes extends up on to body. **DISTRIBUTION** Southern inland Qld. **HABITAT AND HABITS** Small arboreal species found in dry wooded habitats, with a particular association with black wattle and white cedar. Shelters beneath loose bark and inside tree hollows. Nocturnal, mainly feeding on insects and spiders, but will also lick sap and nectar. Members of this genus produce a foul-smelling, sticky substance from their tails when threatened, which can be ejected with great force. If it enters the eyes or open wounds it can cause great discomfort, minor swelling and significant pain. Harmful

Perentie ■ *Varanus giganteus* TL 2.4m

DESCRIPTION Pale brown to dark brown or grey with yellow, cream or white spots, forming transverse bands. Throat is reticulated with black or grey over white, and tail-tip is white or yellow. **DISTRIBUTION** WA across to Qld, including northern SA and NT. **HABITAT AND HABITS** Australia's longest lizard, found in dry habitats including deserts, rocky hills, black soil grassland and coastal heaths. Largely terrestrial, sheltering in burrows

and small caves, and emerging to feed diurnally, predominantly on lizards, snakes and carrion, but will also hunt birds and mammals. Lays 7–13 eggs in a clutch. Large, backwards facing teeth can easily slice through flesh, and strong tail is used like a whip. Venom of this species has not been shown to have an effect on people. Potentially Dangerous

Yellow-spotted Monitor ■ *Varanus panoptes panoptes* TL 1.6m

DESCRIPTION Pale brown to dark brown or grey, with or without yellow, cream or white spots with wide black bands; smaller yellow spots often arranged in transverse

rows. Tail yellow and usually banded with brown to tip. **DISTRIBUTION** From Qld to northern WA, and northern NT; a second smaller subspecies (*V. p. rubidus*) occurs from Pilbara to goldfields in WA. **HABITAT AND HABITS** Found in a huge range of habitat types, thriving almost anywhere other than closed rainforests. Large terrestrial species that shelters in burrows and small caves. Feeds diurnally predominantly on lizards, snakes and carrion, and sea-turtle eggs, but will also hunt birds and mammals. Lays 6–19 eggs in a clutch. Though non-aggressive, the large, backwards facing teeth can easily slice through flesh and the strong tail is used like a whip. Venom of this species has not been shown to have an effect on people. Harmful

Lace Monitor ■ *Varanus varius* TL 2.4m

DESCRIPTION Two distinct colour forms. One is black to dark grey above with yellow to white spots often forming bands. Banded tail. Head and neck can have blue flecking. Other form is known as the 'Bells' phase, and is black with broad yellow or white bands. Tail-tip in both forms is white or yellow. **DISTRIBUTION** Eastern Australia from SA, across Vic and NSW, and north to Cooktown Qld. **HABITAT AND HABITS** Diurnal. Australia's heaviest lizard. Stunning monitor lizard found in many ecosystems, from rainforests to mallee, as well as open woodland, brigalow, grassland and coastal heaths. Large arboreal

species that shelters in burrows, hollowed out trees, termite mounds and caves. Adults predominantly feed on arthropods, carrion, fish, reptiles and birds. Also readily takes food from people at camp grounds and picnic areas. Lays 5–14 eggs in a clutch. Not usually aggressive but can become an issue around places where it is fed by people. The large, backwards facing teeth are extremely sharp and easily slice through flesh, while the strong tail is used like a whip. Venom of this species has not been shown to have an effect on people. Potentially Dangerous

Olive Python ■ *Liasis olivaceus* TL 3m

DESCRIPTION Top of body brown to greyish. Belly white, cream to yellow, without markings. **DISTRIBUTION** Across northern Australia from western Qld to Kimberley WA. Additional subspecies, *L. o. barroni*, occurs only in Pilbara region WA. **HABITAT AND HABITS** Found in open woodland, savannah, swamps, rocky hillsides and river

edges. Occasionally occurs in houses. Nocturnal, but sometimes found basking during the day, particularly during cooler months of the year. Ambush predator whose diet includes large prey such as wallabies, Freshwater Crocodiles, bandicoots, possums and occasionally lizards. Lays 5–12 eggs. A very large individual could theoretically pose a threat to an infant. Harmful

Carpet Python ■ *Morelia spilota variegata* TL 2.9m

DESCRIPTION Extremely variable. Top of body brown, black, grey or yellow, with white to cream or yellow markings that form irregular bands, stripes and blotches. Belly white with grey or black flecks and blotches. Carpet Pythons have a heavy build as they get older, and large individuals routinely have a girth greater than that of a man's arm. **DISTRIBUTION** Much of Australia from Port Macquarie NSW to Kimberley region WA. Only northern NSW and southern Qld individuals attain a very large size. **HABITAT AND HABITS** Occurs in rainforests, forests, open woodland, rocky hills, grassland, savannah, swamps, mangroves and river edges. Commonly found in houses and other properties.

Nocturnal, but often found basking during the day, particularly during cooler months of the year. Ambush predator whose diet includes large prey such as wallabies, brush-turkeys, bandicoots, possums and rats. In urban environments routinely eats pets. Lays 5–56 eggs. Unpredictable disposition. A number of newspaper reports have claimed that Carpet Pythons have attempted to eat young children. While possible, this has not been proven, but it is prudent to take care around these snakes and leave them alone. Harmful

Scrub Python ■ *Simalia amethistina* TL 5m
(Amethystine Python)

DESCRIPTION Top of body brown, with yellowish markings that form irregular bands and blotches. Scrub Pythons have an overall pearlescent sheen that causes refraction, hence the alternate name of Amethystine Python. Belly white to cream without markings. Australia's largest snake. Thin build for a python of its size, with large, regular head scales. **DISTRIBUTION** In Australia, north-east Qld. Also PNG. **HABITAT AND HABITS** Found in rainforests, forests, savannah margins, swamps, mangroves and river edges. Commonly occurs in houses and other properties that it has entered searching for prey.

Nocturnal, but occasionally found basking during the day, particularly during cooler months of the year. Ambush predator whose diet includes larger prey such as wallabies, brush-turkeys, bandicoots, possums and occasionally lizards. Lays 5–12 eggs. Nervous disposition in captivity, but wild individuals are usually quite calm. Extremely powerful snake that can overpower a healthy adult, but this would only be likely ever to occur when someone attempts to handle it. Potentially Dangerous

Common Death Adder ■ *Acanthophis antarcticus* TL 75cm

DESCRIPTION Top of body reddish-brown to charcoal-grey with lighter cross-bands. Lips white with black or dark grey markings. Belly colour similar to that of lighter bands, with darker flecking. Tail terminates in soft spine that is white to yellow or occasionally black. Stocky snake with raised scales above eyes.

DISTRIBUTION From southern WA along southern coast to SA, and through of most of NSW into north-east Qld. **HABITAT AND HABITS** Found in rainforests, forests, grassland and saltbush associations, mallee, heaths and coastal dunes. Nocturnal, but occasionally found basking during the day while in ambush. Settles down in leaf litter and loose soil, and hides in wait for a prospective meal. Tail is usually placed near to head and wiggled to give the appearance of a grub or worm, luring prospective prey items closer. Diet includes rodents, frogs, lizards and occasionally birds. Unlike most Australian snakes, Common Death Adders often do not move away at the approach of a predator, relying on their superb camouflage. They give birth to live young, with 2–32 in a litter. With one of the fastest strikes of any snake, they have highly toxic venom. Death Adder or Polyvalent Antivenom is used to neutralize bites from this species. DANGEROUSLY VENOMOUS

Floodplain Death Adder ■ *Acanthophis hawkei* TL 90cm

DESCRIPTION Top of body sandy-yellow to charcoal-grey with lighter cross-bands. Edges of scales in the bands are much brighter, so that when the snake flattens out indicating that it is scared or threatened, bright colours appear, startling the predator. Lips white with smudged darker markings. Belly colour similar to that of lighter bands with darker flecking. Tail terminates in soft spine that is white to orange or occasionally black. The largest of all death adders.

DISTRIBUTION From inland Qld across Gulf of Carpentaria, to floodplains south of Darwin NT. **HABITAT AND HABITS** Found in blacksoil plains, grassland, floodplains and swamps. Ecology as for the Common Death Adder (see above). Gives birth to live young, with 8–27 in a litter. Death Adder or Polyvalent Antivenom is used to neutralize bites from this species. DANGEROUSLY VENOMOUS

Kimberley Death Adder ■ *Acanthophis lancasteri* TL 65cm

DESCRIPTION Top of body reddish-brown to orange or greyish, with lighter cross-bands. Lips cream with brown to grey markings. Belly colour cream to yellow. Tail terminates in soft spine that is white to black. Stocky snake with raised scales above eyes. **DISTRIBUTION** Kimberley region WA, possibly extending into neighbouring areas of NT. **HABITAT AND HABITS** Occurs in savannah woodland and grassland in

close association with rocky outcrops. Ecology as for the Common Death Adder (see p. 127). This species was recently revised with a description giving it the name *Acanthophis cryptamydros*, but the species name *lancasteri* has since been reinstated. The only record of breeding for the species gave a litter size of 27. Death Adder or Polyvalent Antivenom is used to neutralize bites from this species. DANGEROUSLY VENOMOUS

Northern Death Adder ■ *Acanthophis praelongus* TL 45cm

DESCRIPTION Top of body yellow to reddish-brown to dark grey, with lighter cross-bands. Lips white with black or dark grey markings. Belly colour similar to that of lighter bands with darker flecking. Tail terminates in soft spine that is white to orange, or occasionally black. Stocky snake with the highest raised scales above the eyes of all Australian death

adders. **DISTRIBUTION** Restricted to north-east Qld. **HABITAT AND HABITS** Found in rainforests, forests, grassland and scrubland. Ecology as for the Common Death Adder (see p. 127). Gives birth to live young, with 6–17 in a litter. Death Adder or Polyvalent Antivenom is used to neutralize bites from this species. DANGEROUSLY VENOMOUS

Desert Death Adder ■ *Acanthophis pyyrhus* TL 75cm

DESCRIPTION Top of body orange to reddish-brown with lighter cross-bands. Lips peppered orange-brown with bottom edges white. Belly colour white to cream. Tail terminates in soft spine, which is white or occasionally black. Stocky snake with raised scales above eyes. **DISTRIBUTION** Western and central Australia. **HABITAT AND HABITS** As its name suggests, occurs in both sand-ridge and rocky desert regions and associated adjoining habitats, with a strong preference for spinifex. Ecology as for the Common Death Adder (see p. 127). Gives birth to live young, with 9–14 in a litter. Death Adder or Polyvalent Antivenom is used to neutralize bites from this species. DANGEROUSLY VENOMOUS

Woodland Death Adder ■ *Acanthophis rugosus* TL 75cm

DESCRIPTION Top of body yellow to charcoal-grey with lighter cross-bands. Edges of scales in the bands are much brighter, so when the snake flattens out indicating that it is scared or threatened, bright colours appear, startling a predator. Lips white with black or dark grey markings. Belly colour similar to that of lighter bands with darker flecking. Tail terminates in soft spine that is white to orange or occasionally black. Stocky snake with raised scales above eyes and keeled head-shield. **DISTRIBUTION** Across northern Australia from Kimberley region WA into western Qld. Also PNG. **HABITAT AND HABITS** Found in forests and grassland with rocky soils. Another population of snakes from the Dajarra to Mount Isa area is currently assigned to this species, but further research is needed to determine the status of these snakes. Ecology as for the Common Death Adder (see p. 127). Gives birth to live young, with 6–24 in a litter. Death Adder or Polyvalent Antivenom is used to neutralize bites from this species. DANGEROUSLY VENOMOUS

Pilbara Death Adder ■ *Acanthophis wellsei* TL 55cm

DESCRIPTION Top of body pale brown to orange with lighter cross-bands; another colour phase is red with black bands. Belly colour similar to that of lighter bands with darker flecking. Tail terminates in soft spine that is white or occasionally black. Stocky snake with raised scales above the eyes. **DISTRIBUTION** Pilbara region and Cape Range area WA. **HABITAT AND HABITS** Favours rocky deserts and gorges, and associated adjoining habitats, with a strong preference for spinifex. Ecology as for the Common Death Adder (see p. 127). Gives birth to live young, with 9–20 in a litter. Possibly a species complex. Death Adder or Polyvalent Antivenom is used to neutralize bites from this species. DANGEROUSLY VENOMOUS

Pygmy Copperhead ■ *Austrelaps labialis* TL 75cm

DESCRIPTION Top of body grey-brown to charcoal-grey. Lips boldly marked with black or dark grey on white background. Belly colour cream to yellow with occasional orange flecks. Robust snake with scales that are matt in finish. **DISTRIBUTION** South-east SA in Adelaide Hills and Kangaroo Island. **HABITAT AND HABITS** Occurs in coastal dunes, forests, grassland, swamps and riverine systems. Copperheads also readily exploit human disturbed habitats such as rural areas and suburban backyards. Diurnal, but becomes nocturnal or crepuscular in hot weather. Found under cover, for example beneath logs and rocks, or seen while basking, often inside grass tussocks. Mainly eats frogs and lizards, but will take rodents and snakes. Copperheads are among Australia's most cold-tolerant

group of snakes, often being the last to enter winter torpor and the first to emerge in spring. Males fight slightly differently from other male snakes, with their heads separated from each other – it has been suggested that they may do this because they attempt to eat each other. Produces 3–32 young in a litter. Tiger Snake or Polyvalent Antivenom is used to neutralize bites from this species. DANGEROUSLY VENOMOUS

Highland Copperhead ■ *Austrelaps ramsayi* 1.3m
(Alpine Copperhead)

DESCRIPTION Top of body light brown to charcoal-grey; thin vertebral stripe in some individuals. Lips boldly marked with black or dark grey markings on white background. Belly colour cream to yellow with occasional orange flecks. Copperheads are robust snakes with scales that are matt in finish. **DISTRIBUTION** South-eastern Australia from eastern Vic, up eastern seaboard to Qld border. Distribution becomes patchier in north of range.

HABITAT AND HABITS

Found in forests, grassland, heaths, swamps and riverine systems. Copperheads also readily exploit human disturbed habitats such as rural areas and suburban backyards. Ecology as for the Pygmy Copperhead (see opposite). Produces 5–31 young in a litter. Tiger Snake or Polyvalent Antivenom is used to neutralize bites from this species. DANGEROUSLY VENOMOUS

Lowland Copperhead ■ *Austrelaps superbus* 1.3m
(Superb Snake)

DESCRIPTION Top of body light brown to charcoal-grey. Lateral stripe invariably present that is usually yellow to copper coloured, and often extends up on to the nape, hence the name 'copperhead'. Some individuals also have spots dotted along the body. Lips have dark markings on a white background. These markings do not have a sharp delineation like those of both the Pygmy and Highland Copperheads (see opposite and above). Belly colour cream to yellow with occasional orange flecks. Copperheads are robust snakes with scales that are matt in finish. **DISTRIBUTION** From SA across Vic, Tas, and into southern NSW. Also present on islands of Bass Strait. **HABITAT AND HABITS** Found in forests, grassland, heaths, swamps and riverine systems. Copperheads also readily exploit human disturbed habitats such as rural areas and suburban backyards. Ecology as for the Pygmy Copperhead. Produces 2–32 young in a litter. Tiger Snake or Polyvalent Antivenom is used to neutralize bites from this species. DANGEROUSLY VENOMOUS

Small-eyed Snake ▪ *Cryptophis nigrescens* TL 60cm

DESCRIPTION Top of body grey to jet-black, with black head. Belly colour pink to orange-red. Superficially similar to the Red-bellied Black Snake (see p. 143), but readily differentiated from it by not having black edges to ventral scales. Older individuals can develop a condition called macrocephaly, where the head enlarges past its normal size. **DISTRIBUTION** From Melbourne Vic, up east coast of Australia, to Mossman in north Qld. **HABITAT AND HABITS** Inhabits forests, heaths and scrubland dominated by

rock outcrops. Often occurs in suburban backyards and houses. Nocturnal. Found under cover, for example beneath logs, rocks and bark of fallen logs, or while hunting at night. Diet consists of lizards, frogs and occasionally other snakes. Gives birth to 2–8 live young in a litter. Tiger Snake or Polyvalent Antivenom is used to neutralize bites from this species. DANGEROUSLY VENOMOUS

Lesser Black Whip Snake ▪ *Demansia vestigiata* TL 1.2m

DESCRIPTION Top of body grey to black, with each of the dorsal scales with a darker rear edge, giving snake a variegated appearance. Some individuals have yellow forebody. Belly colour grey. **DISTRIBUTION** From Brisbane Qld, up east coast, and across north

coast of Australia. Also southern PNG. **HABITAT AND HABITS** Inhabits forests, open woodland and grassland. Diurnal, but can be nocturnal in hot weather. Found sheltering under cover, including beneath logs, rocks and debris. Diet consists of lizards, frogs and occasionally other snakes. Egg layer, producing clutch of up to 15 eggs. Was not thought to be a potentially dangerous species until the death of a young man as a result of its bite in PNG in 2007. Polyvalent Antivenom is used to neutralize bites from this species. DANGEROUSLY VENOMOUS

Greater Black Whip Snake ■ *Demansia papuensis* TL 1.4m

DESCRIPTION Top of body grey to black; some individuals are tan to reddish-brown. Belly colour grey. **DISTRIBUTION** From Mackay Qld, up east coast, and across north coast of Australia. Also southern PNG. **HABITAT AND HABITS** Occurs in forests, open woodland and grassland. Diurnal, but can be nocturnal in hot weather. Found under cover, for example beneath logs, rocks and man-made debris. Diet consists of lizards, frogs and occasionally other snakes. Egg layer, producing clutch of 5–13 eggs. Polyvalent Antivenom is used to neutralize bites from this species. Potentially Dangerous

Bardick ■ *Echiopsis curta* TL 70cm

DESCRIPTION Top of body yellow to dark brown or grey. Some individuals are reddish-orange, and some have barring on lips. Belly colour yellowish to cream. **DISTRIBUTION** Disjunct distribution across southern Australia, with three separate populations. One is in south-west WA, another on Eyre Peninsula of SA, and one on in western Vic, south-west NSW and neighbouring SA. **HABITAT AND HABITS** Inhabits dry forests, mallee, heaths and coastal dune assemblages. Nocturnal. Found sheltering under cover, for example beneath logs, in clumps of spinifex and under man-made debris. Diet consists of lizards, frogs, small mammals, and occasionally birds and insects. Gives birth to live young, with 3–14 in a litter. Death Adder or Polyvalent Antivenom is used to neutralize bites from this species if required. Potentially Dangerous

Dunmall's Snake ■ *Glyphodon dunmalli* TL 75cm

DESCRIPTION Top of body dark brown to greyish-black. Occasionally some yellowish flushing on face. Belly colour white to light grey. **DISTRIBUTION** From NSW/Qld

border north around Rockhampton Qld. **HABITAT AND HABITS** Occurs in dry forests, brigalow and scrubland, where it can be seen foraging on the ground or crossing roads at night. Nocturnal. Found sheltering under cover, for example beneath logs and rocks. Diet consists of lizards, reptile eggs and frogs. Produces 5–9 eggs in a clutch. Some authorities place this species in the same genus as the naped snakes *Furina*, to which it is related. No known antivenom is presently recommended for bites from this species. Potentially Dangerous

Brown-headed Snake ■ *Glyphodon tristis* TL 80cm

DESCRIPTION Top of body brown to greyish-black. Conspicuous yellow band across nape that fades to obscurity in old individuals. Head usually lighter brown than body. White skin between scales is prominent, giving a reticulated appearance. Belly colour white to light grey. **DISTRIBUTION** Disjunct distribution, with one population in north Qld, and another in far north-eastern tip of NT. Also PNG. **HABITAT AND HABITS** Inhabits forests,

savannah and scrubland. Nocturnal. Often seen foraging on the forest floor or crossing roads at night. Found sheltering under cover, for example beneath logs and rocks. Diet consists of lizards, reptile eggs and frogs. Produces 6–8 eggs in a clutch. Some authorities place this species in the same genus as the naped snakes *Furina*, to which it is related. Polyvalent Antivenom has been used for bites from this species. Potentially Dangerous

Pale-headed Snake ■ *Hoplocephalus bitorquatus* TL 60cm

DESCRIPTION Top of body silver-grey to black. Conspicuous white band across nape on juvenile and young specimens, which fades to pale saddle in old individuals. Face blotched with black spots. Belly colour grey.
DISTRIBUTION Disjunct distribution, with one population in north Qld, and another further south in southern Qld into NSW. **HABITAT AND HABITS** Occurs in dry forests, open woodland, brigalow and scrubland. Nocturnal. Often seen sitting exposed on trunks of large trees or crossing roads at night. Found sheltering under cover, for example under logs, in tree hollows and beneath bark of standing trees. Diet consists of lizards, frogs and small mammals. Gives birth to 2–17 young. Has toxic venom that has caused serious symptoms rapidly following a bite. Tiger Snake or Polyvalent Antivenom is used to neutralize bites from this species. DANGEROUSLY VENOMOUS

Broad-headed Snake ■ *Hoplocephalus bungaroides* TL 80cm

DESCRIPTION Top of body black with yellow to white flecks that form irregular, thin cross-bands. Belly colour grey. **DISTRIBUTION** Sydney and surrounding areas of NSW. **HABITAT AND HABITS** Inhabits eucalypt forests and heaths with sandstone escarpments and rock exfoliations. Nocturnal, but occasionally found partially exposed during the day, basking. Shelters under cover, for example beneath rocks and behind loose bark of standing trees. Diet consists of lizards and occasionally small mammals. Gives birth to litter of 2–12 young. Endangered species that is threatened by habitat destruction and fragmentation. There are reports that it has been involved in causing a fatality, but they have not been substantiated. Tiger Snake or Polyvalent Antivenom is used to neutralize bites from this species. DANGEROUSLY VENOMOUS

Stephen's Banded Snake ■ *Hoplocephalus stephensii* TL 90cm

DESCRIPTION Top of body dark grey to black. Bands more distinct in young and juvenile individuals than in adults, and unbanded populations are known, which resemble Pale-headed Snakes (see p. 135). Face blotched with black spots. Belly colour grey. **DISTRIBUTION** From Ourimbah NSW to Maryborough Qld. Two isolated populations occur at Eungella and Kroombit Tops Qld. **HABITAT AND HABITS** Found in forests, granite outcrops and scrubland. Nocturnal. Often seen sitting exposed on trunks of large trees or crossing roads at night. Shelters under cover, for example beneath logs, in tree hollows and under bark of standing trees. Diet consists of lizards, frogs and small mammals. Gives birth to litter of 2–17 young. Has toxic venom that has caused serious symptoms rapidly following a bite. Tiger Snake or Polyvalent Antivenom is used to neutralize bites from this species. DANGEROUSLY VENOMOUS

Tiger Snake ■ *Notechis scutatus* TL 1.3m

DESCRIPTION Very variable in both colour and pattern. Top of body can be any shade of brown, grey, black or yellow, with or without cross-bands. Belly colour yellow to grey. **DISTRIBUTION** Southern Qld, through eastern NSW, most of Vic, Tas, southern SA and south-western corner of WA. **HABITAT AND HABITS** Inhabits swamps, forests, grassland,

rainforests, wallum, open woodland, heaths and scrubland. Usually diurnal, but becomes nocturnal during warm weather. Very common in urban areas, where it is the most commonly found snake in backyards and inside houses. Found sheltering under cover, for example in man-made debris, beneath logs and under rocks, or seen basking. Diet consists of lizards, frogs, birds and small mammals. Gives birth to live

(TIGER SNAKE, CONTINUED)

young, with 6–69 to a litter. Until recently Tiger Snakes were regarded as the leading cause of snakebite death in Australia, and even the bites of juveniles have caused very serious symptoms. Tiger Snake or Polyvalent Antivenom is used to neutralize bites from this species. DANGEROUSLY VENOMOUS

Inland Taipan ■ *Oxyuranus microlepidotus* TL 1.7m
(Fierce Snake; Small-scaled Snake; Western Taipan)

DESCRIPTION Very variable in colour. Top of body can be any shade of brown, grey, black or yellow. Goes through seasonal colour change, being much lighter in summer. Head usually shiny black but can be lighter. Belly colour yellow with darker flecking. **DISTRIBUTION** Far western Qld, north-east SA, south-west corner of NT, and into western NSW. There are historical reports of Inland Taipans occurring at the junction of the Darling and Murray Rivers, but despite intensive searching none has been seen in the region for many years. **HABITAT AND HABITS** Inhabits black soil plains, gibber desert, grassland and savannah. Diurnal, but can be crepuscular in hot weather. Found basking on edges of cracks in the soil. Lives in the cracks, and hunts for small mammals in them. This way of life is probably the reason why this species has evolved to be the world's most toxic snake (in laboratory tests conducted with mice). Lays 8–23 eggs. Taipan or Polyvalent Antivenom is used to neutralize bites from this species. DANGEROUSLY VENOMOUS

Coastal Taipan ■ *Oxyuranus scutellatus* TL 2.1m

DESCRIPTION Very variable in colour. Top of body can be any shade of brown, grey, black or yellow, and some individuals have a reddish-orange stripe along the spine that widens towards the rear. Head usually lighter, often cream to white. Belly colour yellow to orange with or without red flecking. **DISTRIBUTION** From Qld/NSW border region across northern coastline into Kimberley region WA. Also islands of PNG. The snakes on PNG and the Torres Strait Islands were thought to be a different subspecies from mainland animals, but there are no genetic or reliable morphological characteristics that separate the two; as a result, subspecies *canni* is no longer recognized. **HABITAT AND HABITS** Occurs in dry forests, grassland, savannah, open woodland, heaths and rainforest verges. Usually

diurnal, but becomes nocturnal during warm weather. Extremely alert and seemingly quite uncommon in southern part of its range; occasionally occurs in backyards and inside houses. Found sheltering under cover, for example in man-made debris, beneath logs and under rocks, or seen while basking. Diet consists of small mammals and rarely birds. Lays 6–22 eggs. Australia's – if not the world's – most dangerous snake, with extremely toxic venom, long fangs and a nervous disposition, and it can be formidable. However, it is much more afraid of people than we are of it and will retreat whenever possible. Taipan or Polyvalent Antivenom is used to neutralize bites from this species. DANGEROUSLY VENOMOUS

Western Desert Taipan ■ *Oxyuranus temporalis* TL 2m

DESCRIPTION Very variable in colour. From the few specimens known, yellow to dark brown. As in Inland Taipans (see p. 137), there is a shift in individual colouration during the seasons, with colouration being lightest in summer. Belly colour yellow to cream, with or without orange flecking. **DISTRIBUTION** Currently known from three localities in remote eastern portion of WA and one locality in southern NT. **HABITAT AND HABITS** Found in deserts and mallee regions with sand ridges, gravel beds, spinifex and shrubs. Currently, almost all individuals have been seen crossing roads during the morning. One was killed behind a roadhouse. Shelter sites used are unknown but thought to be similar to those of other taipans. Diet consists of small mammals. Recently discovered, and not much is known about this interesting species. Lays eggs, but clutch size is currently unknown. Preliminary venom research has found that Taipan Antivenom will provide protection against bites of this species. DANGEROUSLY VENOMOUS

Lake Cronin Snake ■ *Paroplocephalus atriceps* TL 60cm

DESCRIPTION Top of body silver-grey to brown, and head jet-black. Lips barred with white. Belly colour grey. **DISTRIBUTION** Around Lake Cronin WA. **HABITAT AND HABITS** Occurs in dry forests, granite outcrops and adjoining habitats. Found sitting exposed on trunks of large Salmon Gums or crossing roads. Nocturnal. Seen sheltering under cover of, for example, rocks and man-made debris. Diet consists of lizards, frogs and small mammals. Thought to be a live-bearing species. It has toxic venom that has rapidly caused serious symptoms following a bite. Polyvalent Antivenom has been used to neutralize bites from this species. Potentially Dangerous

Little Whip Snake ■ *Parasuta flagellum* TL 40cm

DESCRIPTION Top of body brown to greyish. Head black on top, with pale bar between nostrils and eyes. Belly colour yellow. **DISTRIBUTION** From eastern SA, throughout Vic and into southern NSW. **HABITAT AND HABITS** Occurs in dry forests, open woodland, grassland and scrubland. Nocturnal. Seen crossing roads at night and in swimming pools in urban areas. Found sheltering under cover, for example beneath rocks and logs. Diet

consists of lizards and occasionally frogs. Gives birth to live young, with 2–11 in a litter. For many years this species was thought to be almost harmless, until the bite from an individual in 2007 caused the death of an adult human. It is likely that there was an anaphylactic reaction involved. No known antivenom is presently recommended for bites from this species. Potentially Dangerous

Mulga Snake ■ *Pseudechis australis* TL 2m
(King Brown Snake)

DESCRIPTION Very variable in colour, from black, to pale yellow, to reddish-purple. Southern populations tend to be darker than northern ones. Some individuals have a variegated appearance, with front edges of scales lighter than rear edges. When threatened inflates its body, and appears to brighten. Belly colour yellow to cream, without orange flecking. In some individuals underside of tail is pale orange. **DISTRIBUTION** All over Australia's drier regions throughout almost all of WA, all of NT, most of SA, western NSW and Qld west of Great Dividing Range. **HABITAT AND HABITS** Found in dry woodland, deserts, mallee, agricultural land and heaths. Diurnal to nocturnal, depending on the

temperature. Shelters in burrows, beneath debris, under logs and rocks, and in crevices. Despite the name, a member of the black snake genus. True generalist that will eat other species of snake, lizards, frogs, mammals and even roadkill. Lays 4–23 eggs. Injects the most venom of any Australian snake. Black Snake or Polyvalent Antivenom is used to neutralize bites from this species. DANGEROUSLY VENOMOUS

Spotted Mulga Snake ■ *Pseudechis butleri* TL 1.75m
(Butler's Snake)

DESCRIPTION Black to dark brown with cream to yellow spots. Juveniles grey with shiny black head. Belly colour yellow, and ventral scales can have black spots and black edges.

DISTRIBUTION Southern inland WA. **HABITAT AND HABITS** Found in mulga and acacia woodland, mallee and agricultural land. Diurnal to nocturnal, depending on the temperature. Often shelters in burrows, beneath man-made debris and under logs, and frequently found in abandoned wells. Eats lizards, other snakes and small mammals. Lays 7–14 eggs. Black Snake or Polyvalent Antivenom is used to neutralize bites from this species. DANGEROUSLY VENOMOUS

Collett's Snake ■ *Pseudechis colletti* TL 1.5m
(Down's Tiger Snake)

DESCRIPTION Pale brown to black, with yellow, orange and pink cross-bands and spotting. Belly colour orange with or without dark brown to black flecking. Juveniles much brighter than adults, with orange and jet-black cross-bands. **DISTRIBUTION** Central western Qld. **HABITAT AND HABITS** Found in blacksoil plains and grassland. Diurnal to nocturnal, depending on the temperature. Often shelters in burrows, beneath man-made debris, under logs and rocks, and in crevices formed in cracking soils. Eats lizards, frogs, snakes and small mammals. Lays 7–18 eggs. Black Snake or Polyvalent Antivenom is used to neutralize bites from this species. DANGEROUSLY VENOMOUS

Spotted Black Snake ■ *Pseudechis guttatus* TL 1.8m
(Blue-bellied Black Snake)

DESCRIPTION Very variable in colour, from jet-black to grey and occasionally pale brown. Some individuals flecked with yellow, red, orange or cream. Belly colour grey to yellowish with darker flecking. **DISTRIBUTION** NSW and Qld, almost exclusively west of Great Dividing Range. In some places, such as west Brisbane Qld, reaches over the foothills into the blacksoil valleys, making it through the range. **HABITAT AND HABITS** Found in dry woodland, savannah, grassland, brigalow and agricultural land. Diurnal to nocturnal, depending on the temperature. Often shelters in burrows, beneath man-made debris, under logs and rocks, and in cracks in the soil. Eats small mammals, frogs, lizards and other snakes. Lays 5–16 eggs. Tiger Snake, Black Snake or Polyvalent Antivenom is used to neutralize bites from this species. DANGEROUSLY VENOMOUS

Eastern Pygmy Mulga Snake ■ *Pseudechis pailsei* TL 1.2m

DESCRIPTION Pale yellow to golden brown with magenta markings on nape. When threatened flares neck, looking larger and more dangerous. Belly colour yellow to cream. Narrow head thought to be adapted for squeezing into crevices. **DISTRIBUTION** Along

Selwyn range in western Qld from Winton to Riversleigh. **HABITAT AND HABITS** Found in dry woodland and rocky gorges, and on hillsides. Diurnal to nocturnal, depending on the temperature. Seeks shelter in rock crevices and tree hollows, beneath debris, or under logs and rocks. Eats lizards and frogs, and occasionally small mammals. Lays 5–11 eggs. For many years remained undetected to science, and thought to be a young mulga snake. Black Snake or Polyvalent Antivenom is used to neutralize bites from this species. DANGEROUSLY VENOMOUS

Papuan Black Snake ▪ *Pseudechis papuanus* TL 2.1m

DESCRIPTION Black to dark grey, occasionally flushed with dark red. Belly colour dark grey. **DISTRIBUTION** Saibai and Boigu islands in northern Torres Strait Qld. Also throughout southern PNG. **HABITAT AND HABITS** Found in swamps, forest edges and moist grassland. Mainly diurnal, becoming crepuscular in warm weather. Shelters in burrows or under debris, logs and rocks. Eats frogs and lizards, but has been recorded also eating small mammals and other snakes. Declining over much of its range due to habitat destruction and introduction of the Cane Toad (see p. 122). Lays 7–18 eggs. The most toxic of the black snakes. Black Snake or Polyvalent Antivenom is used to neutralize bites from this species DANGEROUSLY VENOMOUS

Red-bellied Black Snake ▪ *Pseudechis porphyriacus* TL 1.6m
(Common Black Snake)

DESCRIPTION Jet-black above with orange, red or maroon markings along lower flanks. Some individuals marked with white. Head, in particular around snout, can be brown in some places. Belly red and marked with black bands. Tail black beneath. **DISTRIBUTION** East coast from Adelaide Hills SA to around Cooktown in north Qld. **HABITAT AND HABITS** Found in swamps, creeks, forests, rainforests, urban environments and grassland, particularly around waterbodies. Mainly diurnal, but becomes crepuscular in warm weather. Shelters in burrows or under logs, debris and rocks. Eats frogs and lizards, other snakes, fish and occasionally small mammals. Declined in Qld following the introduction of the Cane Toad (see p.122), but now appears to be beginning to recover. Well known to bushwalkers, and individuals can sometimes get used to people and become slow to retreat. Gives birth to 5–23 live young. Tiger Snake, Black Snake or Polyvalent Antivenom is used to neutralize bites from this species. DANGEROUSLY VENOMOUS

Western Pygmy Mulga Snake ■ *Pseudechis weigeli* TL 1.2m

DESCRIPTION Pale yellow to golden-brown or grey, with magenta markings on nape. These darker reticulations can extend all the way down the body. When threatened flares neck, looking larger and more dangerous. Belly colour yellow to cream. Narrow head thought to be adapted for squeezing into crevices. **DISTRIBUTION** Kimberley region WA and Top End NT. **HABITAT AND HABITS** Found in dry woodland, grassland, rocky gorges and hillsides. Diurnal to nocturnal, depending on the temperature. Seeks shelter in rock crevices and tree hollows, beneath debris, or under logs and rocks. Eats lizards and frogs, and occasionally small mammals. Egg layer, with single recorded clutch of seven eggs.

For many years remained undetected to science, and thought to be a young mulga snake. Another species, currently included with the Western Pygmy Mulga Snake, is found in the Top End region of the NT. These snakes are genetically distinct but require further research to determine the validity of the taxa. Black Snake or Polyvalent Antivenom is used to neutralize bites from this species. DANGEROUSLY VENOMOUS

Dugite ■ *Pseudonaja affinis affinis* TL 2m; *P. a. exilis* TL 1.3m; *P. a. tanneri* TL 1.3m
(Spotted Brown Snake)

DESCRIPTION Very variable, from pale cream to almost black. Usually a shade of brown with random dark spots that sometimes coalesce into blotches. Head can be darker or lighter than body. Juvenile and immature individuals have darker head and black band on nape. Belly colour yellow, to cream, to light brown, with orange-red spots and blotches. Both insular subspecies are dark brown to black, above and below. **DISTRIBUTION** Southern WA across southern coast to western edge of Eyre Peninsula SA. The Rottnest

Juvenile

Island Dugite *P. a. exilis* is restricted to Rottnest Island. Tanner's Brown Snake *P. a. tanneri* is restricted to the Boxer and Figure Eight Islands in the Recherche Archipelago, south of WA. **HABITAT AND HABITS** Found in dry woodland, coastal dune associations, heathland, grassland and urban areas. Diurnal to nocturnal, depending on the temperature. Seeks shelter among

(DUGITE, CONTINUED)

vegetation, beneath debris, or under logs and rocks. Eats small mammals and reptiles as adult, but juveniles are almost exclusively reptile predators. Egg layer, producing clutch of 3–31 eggs. Bites from juveniles have killed healthy adult humans. Brown Snake or Polyvalent Antivenom is used to neutralize bites from this species. DANGEROUSLY VENOMOUS

P. a. exilis

P. a. tanneri

Strap-snouted Brown Snake ▪ *Pseudonaja aspidoryncha* TL 1.75m

DESCRIPTION Very variable, from pale cream to dark brown. Usually a shade of brown with or without dark bands. Almost always a black speck on nape. Juvenile and immature individuals have black head and black band on nape. Belly colour yellow with orange-red spots and blotches. **DISTRIBUTION** Inland eastern Australia, including eastern SA, north-west Vic, western NSW and southern west Qld. **HABITAT AND HABITS** Inhabits dry woodland, brigalow, mallee, desert and vicinity of agricultural areas. Diurnal to nocturnal, depending on the temperature. Seeks shelter among vegetation, beneath debris, or under logs and rocks. Eats small mammals and reptiles as adult; juveniles are almost exclusively reptile predators. Egg layer, producing clutch of 13 eggs. Bites from juveniles have killed healthy adults. Brown Snake or Polyvalent Antivenom is used to neutralize bites from this species. DANGEROUSLY VENOMOUS

Speckled Brown Snake ■ *Pseudonaja guttata* TL 1.2m

DESCRIPTION Very variable. Usually a shade of brown with or without dark bands; many individuals are flecked with black. Juveniles and immature individuals have black head and black band on nape. Belly colour yellow with orange-red spots and blotches. **DISTRIBUTION** Inland northern Qld and NT. **HABITAT AND HABITS** Found in

grassland and on black soil plains. Diurnal to nocturnal, depending on the temperature. Seeks shelter among vegetation, beneath debris and in soil cracks. Eats frogs, lizards and occasionally small mammals, although juveniles are almost exclusive reptile and frog predators. Lays 3–17 eggs. Brown Snake or Polyvalent Antivenom is used to neutralize bites from this species. DANGEROUSLY VENOMOUS

Peninsula Brown Snake ■ *Pseudonaja inframacula* TL 1.6m

DESCRIPTION Very variable, from yellow-brown to purplish black. Most individuals flecked with black, sometimes coalescing into blotches. Juveniles and immature individuals have black head and black band on nape. Belly colour grey. **DISTRIBUTION** Southern SA, on Yorke and Eyre Peninsulas. Isolated population on Nullarbor Plain WA. **HABITAT AND HABITS** Found in dry woodland, heaths, grassland, coastal dunes and around agricultural areas. Diurnal to nocturnal, depending on the temperature. Seeks shelter among vegetation, beneath debris, or

under logs and rocks. Eats small mammals, frogs and reptiles as adult, but juveniles are almost exclusive reptile predators. Egg layer, with a clutch of about 12 eggs. Brown Snake or Polyvalent Antivenom is used to neutralize bites from this species. DANGEROUSLY VENOMOUS

Ingram's Brown Snake ■ *Pseudonaja ingrami* TL 2m

DESCRIPTION Very variable, from pale yellow, orange, reddish-brown or dark brown, to black. Juvenile and immature individuals have black head marking. Belly colour yellow with orange-red spots that are arranged in straight lines down body. **DISTRIBUTION** Inland northern Qld and NT. **HABITAT AND HABITS** Found in grassland and on black soil plains. Diurnal to nocturnal, depending on the temperature. Seeks shelter in soil cracks, among vegetation, and beneath logs and rocks. Eats small mammals and reptiles as adult, but juveniles are almost exclusive reptile predators. Lays clutch of 5–18 eggs. Brown Snake or Polyvalent Antivenom is used to neutralize bites from this species. DANGEROUSLY VENOMOUS

Western Brown Snake ■ *Pseudonaja mengdeni* TL 1.4m

DESCRIPTION Very variable, from pale cream to dark brown. Many colour morphs, including banded individuals, snakes marked with herringbone pattern, and bright orange individuals with jet-black head and neck. Juveniles and immature individuals have black head and black band on nape, while others are completely banded. Belly colour yellow to creamish with orange-red spots and blotches. **DISTRIBUTION** Much of arid and west Australia, including SA, western NSW, south-west Qld, arid regions of NT, and most of WA including Kimberley region. **HABITAT AND HABITS** Found in dry woodland, mallee, grassland and deserts, and around agricultural areas. Diurnal to nocturnal, depending on the temperature. Seeks shelter among vegetation, beneath man-made debris, or under logs and rocks. Eats small mammals and reptiles as adult, but juveniles are almost exclusively reptile predators. Lays 7–22 eggs. Bites from juveniles have killed healthy adult humans. Brown Snake or Polyvalent Antivenom is used to neutralize bites from this species. DANGEROUSLY VENOMOUS

Ringed Brown Snake ▪ *Pseudonaja modesta* TL 60cm
(Five-ringed Snake)

DESCRIPTION Grey to reddish-brown, with 4–11 black cross-bands that fade to obscurity with age. Belly colour yellow with orange-red spots and blotches. **DISTRIBUTION** Arid central Australia to western WA coastline. **HABITAT AND HABITS** Found in dry woodland, mallee, grassland and deserts, and around agricultural areas. Diurnal to

nocturnal, depending on the temperature. Seeks shelter among vegetation, beneath man-made debris, or under logs and rocks. Feeds on reptiles. Lays 7–20 eggs. Recent genetic work has shown that this snake may not be a type of brown snake, and is perhaps more than one species. Brown Snake or Polyvalent Antivenom is used to neutralize bites from this species. Potentially Dangerous

Northern Brown Snake ▪ *Pseudonaja nuchalis* TL 1.4m

DESCRIPTION Very variable, from pale brown to gold or dark brown. Some individuals have dark bands, while others have dark nape-band. Usually a black speck on nape. Juveniles and immature individuals have black head and black band on nape, and are sometimes completely banded. Belly colour yellow with orange-red spots and blotches. **DISTRIBUTION** Across northern Australia, west of Great Dividing Range into Kimberleys WA. **HABITAT AND HABITS** Found in dry woodland, tropical savannah, grassland, rocky outcrops and deserts, and around agricultural areas. Diurnal to nocturnal,

depending on the temperature. Seeks shelter among vegetation, beneath man-made debris, in burrows, or under logs and rocks. Eats small mammals and reptiles as adult, but juveniles are almost exclusively reptile predators. Lays 8–16 eggs. Bites from juveniles have killed healthy adult humans. Brown Snake or Polyvalent Antivenom is used to neutralize bites from this species. DANGEROUSLY VENOMOUS

Eastern Brown Snake ■ *Pseudonaja textilis* TL 1.5m
(Common Brown Snake)

DESCRIPTION Very variable, from pale cream to black. Usually a shade of brown with or without dark bands. Juveniles and immature individuals have black head and black band on nape, and some are completely banded. Belly colour yellow to cream, with orange-red spots and blotches; sometimes they are marked with grey. **DISTRIBUTION** Eastern Australia over all of NSW, most of Vic and Qld, and south eastern SA. Isolated populations in NT around Alice Springs and Victoria River district WA, and also in southern PNG. **HABITAT AND HABITS** Found in dry woodland, brigalow, mallee, grassland and deserts, and around agricultural areas. Diurnal to nocturnal, depending on the temperature. Seeks shelter among vegetation, beneath man-made debris, or under logs and rocks. Eats small mammals and reptiles as adult, but juveniles are almost exclusively reptile predators. Lays 6–28 eggs. The Alice Springs snakes are genetically similar to the southern Papuan snakes and may prove to be distinct. Eastern Brown Snakes have killed more people than any other Australian snake, and bites from juveniles have killed healthy adult humans. Brown Snake or Polyvalent Antivenom is used to neutralize bites from this species. DANGEROUSLY VENOMOUS

Juvenile

Curl Snake ■ *Suta suta* TL 60cm
(Myall Snake)

DESCRIPTION Top of body yellow-brown to dark brown or grey. Some individuals have dark edges to posterior of scales, giving a reticulated appearance. Head usually darker than rest of body, but this fades with maturity. Some individuals have light barring on lips and broken yellow stripe along sides of head. Belly colour white to cream. **DISTRIBUTION** Over most of arid Qld, NSW, northern Vic, most of SA and NT. An isolated population occurs near Lake Argyle WA. **HABITAT AND HABITS** Occurs in dry forests, mallee,

heaths and deserts. Found actively hunting or crossing roads at night. Nocturnal. Shelters under cover, for example beneath logs, rocks or man-made debris, or in deep soil cracks, and has been located in trees hunting small dragon lizards. Diet consists of lizards, frogs and small mammals. Produces 1–9 live young. Polyvalent Antivenom is used to neutralize bites from this species if required. Potentially Dangerous

Rough-scaled Snake ■ *Tropidechis carinatus* 90cm
(Clarence River Snake)

DESCRIPTION Top of body brown to grey, occasionally with greenish wash. Some individuals are flecked with black. Northern animals can be completely banded. Belly colour yellowish-green. **DISTRIBUTION** Occurs in two populations. Southern population found from Gosford NSW to Fraser Island Qld, northern population in wet tropics region of north Qld. **HABITAT AND HABITS** Inhabits closed forests, wallum swamps and wetlands. Often found sitting in vegetation or crossing roads at night. Nocturnal,

but basks in the morning. Shelters under cover, for example beneath logs, in tree hollows, or under rocks or man-made objects. Diet consists of small mammals, frogs, tadpoles, small lizards and occasionally birds. Gives birth to litter of 5–19 young. Very toxic venom that has caused death in humans extremely rapidly. Tiger Snake or Polyvalent Antivenom is used to neutralize bites from this species. DANGEROUSLY VENOMOUS

Olive Sea Snake ■ *Aipysurus laevis* TL 1.7m

DESCRIPTION Top of body brown to white with darker and lighter blotches and specks. Some individuals are white with a golden head. Belly colour similar to upper surface.

DISTRIBUTION Waters around Australia, north of Sydney NSW, to Exmouth WA. **HABITAT AND HABITS** Lives in waters over both coral reefs and rocky areas. Diurnal. Deepest diving of all sea snakes, recorded down to 133m depth. Known to eat wide range of small fish, fish eggs, and occasionally prawns and crabs. Gives birth to litter of 1–5 young. Sea Snake Antivenom is expected to neutralize bites from this species. DANGEROUSLY VENOMOUS

Spine-bellied Sea Snake ■ *Hydrophis curtus* TL 1m

DESCRIPTION Top of body pale brown to yellow above, with wide cream to white bands that fade with age. Juveniles white or cream with grey bands. Belly colour similar to upper surface. **DISTRIBUTION** North of Gladstone Qld to Kimberleys WA. Also occurs across Indian Ocean into Arabian Gulf. **HABITAT AND HABITS** Lives in waters over coral reefs, rocky areas and sandy estuaries. Mainly nocturnal. Sexually dimorphic, with males developing spine-like projections on lower scales. Known to eat wide variety of fish. Gives birth to 1–15 live young. Sea Snake Antivenom is used to neutralize bites from this species. DANGEROUSLY VENOMOUS

Elegant Sea Snake ▪ *Hydrophis elegans* TL 1.8m

DESCRIPTION Top of body cream to white with grey to black bands. Belly colour similar to upper surface. Juveniles much brighter than adults, with clear black markings. **DISTRIBUTION** Waters around Australia, north of Sydney NSW, to Exmouth WA.

HABITAT AND HABITS Lives in waters over both coral reefs and rocky areas. Nocturnal. One of the longest of the sea snakes. Known to eat small eels and hole-dwelling fish. Gives birth to 3–30 live young. Sea Snake Antivenom is expected to neutralize bites from this species. DANGEROUSLY VENOMOUS

Horned Sea Snake ▪ *Hydrophis peronii* TL 90cm

DESCRIPTION Top of body brown to grey with whitish blotches and markings; some individuals are banded while others are plain. Belly colour similar to upper surface. Head strongly keeled, with raised scales above eyes. **DISTRIBUTION** Waters around Australia, north of Brisbane Qld. **HABITAT AND HABITS** Lives in waters over both coral reefs and rocky areas, as well as over mudflats and seagrass beds. Nocturnal. Known to eat gobies and other small fish. Gives birth to 1–8 live young. Sea Snake Antivenom is expected to neutralize bites from this species. DANGEROUSLY VENOMOUS

The head is strongly keeled, with raised scales

Yellow-bellied Sea Snake ■ *Hydrophis platura* TL 1m

DESCRIPTION Top of body black to dark brown. Bottom half of snake yellow or white. Some individuals are completely yellow. **DISTRIBUTION** Waters around Australia. Also occurs across Indian Ocean into Arabian Gulf, and to coastline of South America. **HABITAT AND HABITS** Lives in open water. The world's most widespread snake. Mainly diurnal. Uses an ingenious feeding strategy. Sits motionless in the water and waits for small fish to gather around its tail, then rapidly swims backwards, leaving the cluster of fish surrounding its head; striking sideways, it then captures its meal. This strategy has led to an evolutionary change in the species, in which the small fangs have been pushed much further back in the mouth in comparison to its relatives. Gives birth to 1–6 live young. Sea Snake Antivenom is used to neutralize bites from this species. DANGEROUSLY VENOMOUS

Australian Beaked Sea Snake ■ *Hydrophis zweifeli* TL 90cm

DESCRIPTION Top of body white or cream, with grey bands that fade as a snake ages. Juveniles white or cream with grey to black bands. Belly colour similar to upper surface, but lighter. **DISTRIBUTION** Waters around Australia, north of Brisbane Qld around to about Darwin NT. Also PNG. **HABITAT AND HABITS** Lives in waters over sandy inlets and in estuaries, and occasionally found well upriver in fresh water. Mainly nocturnal. Known to eat catfishes, pufferfishes and occasionally prawns. Loose skin beneath jaw may be an adaption to enable the snake to eat larger prey. Gives birth to 1–33 live young. Sea Snake Antivenom is used to neutralize bites from this species. DANGEROUSLY VENOMOUS

Stokes's Sea Snake ■ *Hydrophis stokesii* TL 1.4m

DESCRIPTION Top of body white or cream, with grey to black spots that fade as a snake ages. Juveniles white or cream with grey to black blotches. Belly colour similar to upper surface, but lighter. Very heavy build. **DISTRIBUTION** Waters around Australia, north of Sydney NSW to Exmouth WA. Also occurs across Indian Ocean to Arabian Gulf.

HABITAT AND HABITS Lives in waters over sandy inlets, in estuaries and on coral reefs. Mainly nocturnal. Known to eat frogfishes and stonefishes. Gives birth to 1–14 live young. Sea Snake Antivenom is used to neutralize bites from this species. DANGEROUSLY VENOMOUS

Freshwater Crocodile ■ *Crocodylus johnstoni* TL 2m
(Johnston's Crocodile)

DESCRIPTION Top of body greenish-yellow to brown or greyish with black spotting and black cross-bands. The spots fade as the animals age. Juveniles yellow-brown to greenish above, with grey to black or brown blotches. Belly colour white to cream. **DISTRIBUTION** From Burdekin river catchment in eastern Qld across into Kimberleys WA. **HABITAT AND HABITS** As their name suggests, Freshwater Crocodiles mostly inhabit freshwater bodies, and they penetrate much further inland than Saltwater Crocodiles (see opposite). Found in swamps, rivers and dams. Mainly nocturnal, but often seen basking on the banks of waterways. Fish eaters, but also eat other reptiles, frogs and small mammals. They lay 4–20 eggs. Potentially Dangerous

Saltwater Crocodile ■ *Crocodylus porosus* TL 4m
(Estuarine Crocodile)

DESCRIPTION Top of body greenish-yellow to brown or greyish with black spotting. The spots fade as the animals age. Belly colour white to cream. **DISTRIBUTION** Waters around northern half of Australia, from Fraser Island Qld to about Broome WA. Also occurs throughout Southeast Asia and neighbouring Pacific. **HABITAT AND HABITS** Lives in waterways in wide variety of habitats, including rainforests, mangroves, swamps, beaches and estuaries, in fresh, brackish and salt waters; often seen far out at sea. Mainly nocturnal, but often seen basking on the banks of waterways. Eats fish, birds and other reptiles, but mainly consumes mammals, including people if they get too close to the water. These ambush predators are the largest of all living reptiles. Female lays 40–60 eggs, and vigorously stands guard over her nest and over the hatchlings for the first few months after hatching. DANGEROUS

Southern Cassowary ■ *Casuarius casuarius* TL 1.8m

DESCRIPTION Adults purplish-black with naked, bright blue head and neck. Back of neck bright red. Pair of bright red wattles around base of neck. Juveniles striped with black and white when first hatched, but after a few weeks become dull brown. Top of head has a prominent casque, the function of which is up for debate, but it may be used for getting through dense vegetation and also for defence. **DISTRIBUTION** North-east Qld from Paluma to Pascoe River. Also PNG and Indonesia. **HABITAT AND HABITS** Diurnal, flightless,

and restricted to rainforests and rainforest verges, due to habitat loss and death from collisions with vehicle. Fruit specialist, eating the dropped fruits of plants in rainforests, and spreading the seeds away from fruiting trees. Will take food from trees if it can reach it. Innermost toe has a sharp, dagger-like claw, which this powerful bird uses to defend itself or its offspring if it feels threatened. This action has resulted in human fatalities. DANGEROUS

Emu ■ *Dromaius novaehollandiae* TL 2m

DESCRIPTION Adults grey-brown with black streaks. Head and most of neck covered in thin black feathers. Sides of face and the neck are plain with pale blue to bluish-grey streak.

Juveniles striped with black and white when first hatched, but after a few weeks become dull brown. **DISTRIBUTION** Over almost all of Australia, except for small area near state intersection of WA, NT and SA. **HABITAT AND HABITS** Diurnal, flightless, and widespread across habitats from desert margins to alpine meadows. Not found in dense forests. Varied diet that includes plant sprouts, fruits and insects. This large bird has attacked people who have got too close to its nest or young. Its greatest threat to humans is as a cause of car accidents. Potentially Dangerous

Australian Magpie ■ *Gymnorhina tibicen* TL 40cm

DESCRIPTION Stunning black-and-white bird that has some variation across its range. Southern populations have white backs, western birds have mottled backs, while the rest have black backs. Less contrast in young birds than in adults, with grey replacing black and white area being darker. Eyes reddish-brown. Pale beak tipped with black. In females, back of neck is greyish. **DISTRIBUTION** Over most of Australia, except northern tip of Cape York Peninsula Qld, Top End region NT, through Great Sandy Desert WA and south-western Tas. Also found in PNG and introduced to NZ. **HABITAT AND HABITS** Occurs in urban environments, open woodland and grassland. Diurnal. Feeds on invertebrates, small lizards and frogs. Habituates to people, often becoming tame and taking pieces of meat and bread from them. To protect their eggs and young, magpies dive bomb a perceived threat or swoop at it. This action, along with the sharp beak, has resulted in people being hit by cars as they tried to evade the swooping birds, while others have lost an eye. Many people wear helmets with spike-like projections to deter attacks, but whether this works is not proven. Harmful

Platypus ■ *Ornithorhynchus anatinus* TL 63cm

DESCRIPTION Top of body dark brown with dark grey-brown bill. Belly lighter brown to silver. Tail greatly compressed and paddle-like. Webbed feet. Small white spurs on hindlimbs of males. **DISTRIBUTION** Eastern Australia from northern Qld to Tas. **HABITAT AND HABITS** Found in wide variety of freshwater habitats, including river creeks, pools and small lakes. Cathemeral. During the day mainly rests in a burrow constructed in the bank of a watercourse. Feeds on invertebrates dug out of the waterway substrate, finding them by using its sensitive bill, which can detect the electrical impulses of its prey as it moves below the surface. Platypuses lay two eggs, and when they hatch the female suckles the young for 4–5 months. Venom is delivered via the spurs on the hindlegs, and causes significant swelling and pain in humans that is not counteracted by powerful drugs such as morphine. It is only produced by males during the breeding season in August–October. Venomous

Eastern Grey Kangaroo ■ *Macropus giganteus* TL 2.4m

DESCRIPTION Top of body grey to dark grey-brown. Underside lighter, from white to cream. Hands and ends of feet usually dark grey to black. Males can attain twice the size of females. **DISTRIBUTION** Widespread from Cooktown Qld, south into eastern Tas, across to south-eastern SA.

HABITAT AND HABITS Found in wide variety of habitats such as grassland, open woodland, agricultural areas, brigalow, forest verges and mallee. Cathemeral. Most commonly seen in close proximity to people. Wildlife carers have been killed by 'tame' individuals. Their proximity to people also causes many serious, occasionally fatal, car accidents every year. Potentially Dangerous

Western Grey Kangaroo ■ *Macropus fuliginosus* TL 2.2m

DESCRIPTION Top of body dark brown to grey-brown. Underside lighter, from light grey to cream. Blackish muzzle. Hands and ends of feet usually dark grey to black. Males can attain twice the size of females. **DISTRIBUTION** Widespread from Shark Bay WA, east to north-west Vic, and north through western NSW into Paroo River Drainage of southern central Qld. **HABITAT AND HABITS** Found in wide variety of habitats such as grassland, open woodland, agricultural areas, brigalow, forest verges and mallee. Cathemeral. Potentially Dangerous

Red Kangaroo ■ *Osphranter rufus* TL 2.4m

DESCRIPTION Top of body in males pale red-brown or rust in colour that gradually becomes beige to cream on underside. Females usually grey-blue in colour, becoming beige to cream on underside. Females in central populations usually similar in colour to males. Long eyelashes compared with those of other large kangaroos. Muzzle usually grey with dark-edged broad white stripe extending from lips towards eyes. Males can attain twice the size of females. The largest living marsupial. **DISTRIBUTION** Widespread from north of Shark Bay WA, east to western slopes of Great Dividing Range NSW. **HABITAT AND HABITS** Found in arid to semi-arid habitats, including deserts, mulga, open woodland, grassland and agricultural areas. Cathemeral. Their large size, long, dagger-like toes and speed make them dangerous when upset. Potentially Dangerous

Pig ■ *Sus scrofa* TL 1.5mm

DESCRIPTION Very varied, from white to black with any manner of tones of yellow-red and brown between. Most adults are black or dark brown, while juveniles are usually mottled or striped. **DISTRIBUTION** Introduced, and occurs over most of eastern Australia. Scattered around northern Australia, including Top End NT and Kimberley region WA. **HABITAT AND HABITS** Found across most habitats, with a preference for wetter areas. Cathemeral. Pigs dig out wallows in moist ground to form mud baths. This trait, along with their feeding behaviour, has caused significant damage to many ecosystems across Australia. They are true omnivores, feeding on small animals, carrion, tubers, vegetation, fruits and succulents. Humans can be at risk if they inadvertently get too close to a sow with young. Many people have also been injured when hunting this species for sport. Potentially Dangerous

Swamp Buffalo ■ *Bubalus bubalis* TL 3m

DESCRIPTION Dark grey to blackish, and legs and hindquarters may be flushed with brown. Large horns on both sexes. **DISTRIBUTION** Introduced. Only found in Top End NT. **HABITAT AND HABITS** Mainly found in and around water. Lives in flooded

grassland and flooded woodland. Cathemeral, sheltering away from the heat of the day in wallows or shade of trees. Solitary for most of the year, but occasionally lives in small, same-sex groups. Mainly eats aquatic vegetation, but in the dry season when food is less abundant it will graze. Humans can be charged when they do not give them a wide berth. Females when looking after calves can be fiercely defensive. Potentially Dangerous

Cattle ■ *Bos taurus* TL 2.5m

DESCRIPTION Dark grey to blackish, and legs and hindquarters may be flushed with brown. Large horns on both sexes. **DISTRIBUTION** Introduced. Found across Australia. **HABITAT AND HABITS** Occurs in most habitats, from rainforest margins to almost arid environments. Wild individuals usually more prevalent in isolated, difficult-to-access areas. Cathemeral, sheltering away from the heat of the day in wallows or shade of trees, especially in northern Australia. Lives either in small herds or solitarily, depending on food availability and reproductive status. Grazes mainly on grasses, but will browse on shrubs on occasion. Humans can be charged when they do not give them a wide berth; responsible for car accidents on unfenced roads. Potentially Dangerous

Leopard Seal ■ *Hydrurga leptonyx* TL 3.3m

DESCRIPTION Dark grey with white or silver-coloured mottling above. Belly white with grey to black spots and flecking. Large head and teeth. **DISTRIBUTION** Found mainly around Antarctica, but travels into waters along southern Australian coastline. **HABITAT AND HABITS** Unlike most seals, this species is solitary and nomadic, and is regularly seen out in open water far from land. Turns up occasionally on beaches, resting. Known to prey on wide variety of animals, including smaller seals, penguins, fish, cephalopods and seabirds. Breeds on Antarctic pack ice. Powerful foreflippers can propel it at up to 40km an hour in the water, allowing for an explosive attack. Has been recorded stalking people, and this behaviour has led to at least one death, when a marine biologist was taken while diving. Potentially Dangerous

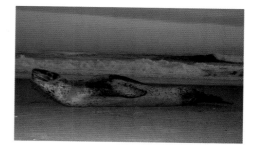

Dingo ■ *Canis lupus dingo* TL 90cm

DESCRIPTION Top of body sandy-yellow to red-brown. Occasionally dark brown to black. Underside lighter, from white to tan. Hybrids with Domestic Dogs are very common but cannot be accurately distinguished visually. **DISTRIBUTION** Historically, found across mainland Australia. Now occurs across northern Australia, north-west SA, and down east coast to Gippsland region Vic. Distribution somewhat restricted due to man-made Dingo fence constructed to keep Dingoes out of much of agricultural land in SA, southern Qld and NSW. **HABITAT AND HABITS** Found in wide variety of habitats, from cool mountainous forests to deserts. Cathemeral. Opportunistic hunters and scavengers, Dingoes hunt in small packs and work together to take down large prey. They will take sheep and young cattle, which has led to them being persecuted. Attacks on people are relatively common in areas where feeding (illegally) by the public occurs. Attacks have proved fatal. Potentially Dangerous

Cat ■ *Felis catus* TL 55cm

DESCRIPTION Very varied, from white to black with any manner of tones of yellow-red and brown between. Most individuals are mottled or striped. **DISTRIBUTION** Introduced to Australia from Europe, and found in all states and territories. **HABITAT AND HABITS** Cats are superbly designed predators that are adaptable to any Australian conditions.

This adaptability, along with the lack of naturally occurring cat predators in Australia, has contributed to the devastation to wildlife caused by this animal. Essentially nocturnal, during the day cats rest in a den that is usually a burrow, small cave or hole in a tree. They are opportunistic hunters, and their diet is largely influenced by the habitat they are subjected to. There is no difference between a 'feral' cat and a typical pet cat. Both present a risk of disease transmission to people by carrying parasites that cause toxoplasmosis and sarcosporidiosis. The common fungal infection called ringworm, and roundworm (a parasite), can also be transmitted from cats to humans. Harmful

Black Flying-fox ■ *Pteropus alecto* TL 25cm

DESCRIPTION Body covered with charcoal-grey to black fur. Wings charcoal-black. Lower legs unfurred. In some individuals back of neck has rust-coloured tint.

DISTRIBUTION Widespread from north of Shark Bay WA, across northern Australia, to Bowraville NSW. **HABITAT AND HABITS** Forms large camps in mangroves, rainforests, and paperbark and wet sclerophyll forests. Nocturnal, roosting during the day. Travels up to 50km from camp each night in search of food. Mainly eats nectar from blooming eucalypts, melaleucas and other native trees. Also eats fruits such as mangoes, mandarins, apples and other produce. Can carry the deadly Lyssavirus (ABLV), as well as Hendra virus. Potentially Dangerous

Spectacled Flying-fox
▪ *Pteropus conspicillatus* TL 23cm

DESCRIPTION Body covered with charcoal-grey to black fur. Fur surrounding eyes pale yellow to straw in colour. Wings charcoal-black. Lower legs unfurred. Back of neck straw coloured. **DISTRIBUTION** Found north of Tully Qld, north on to islands of Torres Strait Qld. **HABITAT AND HABITS** Forms large camps in tall rainforests, and paperbark and wet sclerophyll forest. Nocturnal, roosting during the day. Mainly eats nectar from blooming eucalypts, melaleucas and other native trees. Known to be aggressive towards each other around their favourite food sources. Can carry the deadly Lyssavirus (ABLV), as well as Hendra virus. Potentially Dangerous

Grey-headed Flying-fox ▪ *Pteropus poliocephalus* TL 26cm

DESCRIPTION Body covered with grey fur. Head slightly lighter in colour. Wide band of orange-coloured fur covering neck. Wings charcoal-black. Lower legs lightly furred. The largest Australian fruit bat species. **DISTRIBUTION** From Gladstone Qld, south to Melbourne Vic. **HABITAT AND HABITS** Forms large camps in thick forests; in urban environments utilizes areas like botanical gardens. Nocturnal, roosting during the day. Mainly eats nectar from blooming eucalypts, melaleucas and other native trees. Can carry the deadly Lyssavirus (ABLV), as well as Hendra virus. Potentially Dangerous

Little Red Flying-fox ■ *Pteropus scapulatus* TL 19cm

DESCRIPTION Body covered with reddish-brown fur. Head slightly lighter in colour. In some individuals there is a yellowish-brown flush on the neck. Wings reddish-brown. Lower legs lightly furred. The smallest Australian fruit bat species. **DISTRIBUTION** From Shark Bay WA, across northern Australia, and down east coast into north-western Vic. **HABITAT AND HABITS** Forms large camps in thick forests; in urban environments utilizes areas like botanical gardens. Often shares camps with other species of fruit bat. Nocturnal, roosting during the day. Mainly eats nectar from blooming eucalypts, melaleucas and other native trees. Routinely feeds on cultivated fruit trees. During dry conditions skims the surfaces of waterbodies and drinks from water collected on fur. Can carry the deadly Lyssavirus (ABLV), as well as Hendra virus. Potentially Dangerous

Yellow-bellied Sheath-tailed Bat ■ *Saccolaimus flaviventris* TL 80mm

DESCRIPTION Upperparts of body covered with black fur. Underside yellow to whitish.

Head flattened in appearance, with nose and lips pinkish in colour. Some individuals have yellowish-brown flush on neck. Wings reddish-brown. Lower legs lightly furred. **DISTRIBUTION** Found over most of Australia; absent from south-west and Tas. **HABITAT AND HABITS** Lives in wide variety of habitats, from deserts to rainforests. Roosts in tree hollows in small groups of up to 60 individuals. Also known to roost in burrows of other mammals. Emerges at dusk to hunt its prey of insects. Can carry the deadly Lyssavirus (ABLV). Potentially Dangerous

Horse ■ *Equus caballus* TL 2.1m

DESCRIPTION Usually brown to black or a mixture of both colours, sometimes with white flashes on feet and around muzzle. Domestic horses occasionally breed with wild feral horses, introducing many other colour combinations. **DISTRIBUTION** Introduced, and found in scattered populations around Australia. **HABITAT AND HABITS** Australia's most dangerous animal. More deaths are due to this species than any other because of its contact with people in the agricultural and recreational horse industries. Found mainly in open habitats such as open woodland and grassland. Has caused massive damage to fragile ecosystems in the alpine region. Mainly eats grass but will occasionally browse on shrubs and fruits. Requires fresh water at least every few days, and has been noted digging in dry creek beds to obtain water. Usually lives in small groups of up to 10 individuals. DANGEROUS

House Mouse ■ *Mus musculus* TL 9cm (plus tail 10cm)

DESCRIPTION Pale brown to blackish, with whitish or pale brown belly. Tail mostly naked, with circular annulations (rows) of scales. **DISTRIBUTION** Introduced, most likely from Mediterranean. Found in suitable habitats throughout Australia, often in close association with humans. **HABITAT AND HABITS** Nocturnal, in a range of environments, including urban and natural habitats. Its adaptability has allowed it to thrive. Usually lives in cracks or in a complex network of underground tunnels, often constructed under cover such as debris. Food typically consists of seeds and other plant material, small invertebrates and human food. Can spread diseases through either direct contact with urine or faeces, or through fleas and ticks. More than 35 diseases have been isolated, including Salmonella, Lymphoctyic Choriomeningitis, Weil's disease and Murine typhus. Harmful

Brown Rat ■ *Rattus norvegicus* TL 25cm (plus tail 18cm)
(Sewer Rat)

DESCRIPTION Extremely varied, from brown to grey-brown. Lighter coloured underneath. Ears and tail shorter than those of the Black Rat (see below). **DISTRIBUTION** Introduced, most likely with arrival of Europeans in 18th century. Usually found around ports and human habitation, with strongholds around cities. Origin is thought to be Caspian region of eastern Europe. **HABITAT AND HABITS** Nocturnal, living on the ground and beneath debris. Will dig burrow complexes. Does not usually climb. Feeds on animals including insects, mice, birds, eggs and lizards. Also eats grain, seed and scraps. Readily bites if provoked. Can spread diseases, such as Salmonella, Weil's disease, Murine typhus and rat bite fever. There has been a case in Australia where this species has killed a sleeping infant. Harmful

Black Rat ■ *Rattus rattus* TL 22cm (plus tail 20cm)
(Ship Rat)

DESCRIPTION Extremely varied, from white to black but usually slate-grey. Underneath of body white to cream. Ears and tail much longer than those of the Brown Rat (see above). **DISTRIBUTION** Introduced, most likely with arrival of Europeans in 18th century. Usually occurs around human habitation, with strongholds around cities, but also found well away from human settlements. Thought to have originated in Middle

East. **HABITAT AND HABITS** Nocturnal, living above the ground. Often forms nests in roofs and trees. Feeds on animals including insects, mice, birds, eggs and lizards. Will also eat grain, seed and scraps. Not as aggressive as the Brown Rat but will defend itself if provoked. Can spread diseases through contact with urine or faeces, or through bites from fleas and ticks. Rats are known factors in causing damage to ecosystems in many parts of the world. Harmful

FURTHER INFORMATION

WEBSITES

Animal Diversity Web: www.animaldiversity.org
Australian Faunal Directory: www.biodiversity.org.au
Australian Marine Stinger Advisory Service: www.stingeradvisor.com
Australian Museum: www.australianmuseum.net.au
Australian Society of Clinical Immunology and Allergy: www.allergy.org.au
Australian Wildlife Conservancy: www.australianwildlife.org
Australia's Wildlife: www.australiaswildlife.com
Coffs Harbour Butterfly House: www.lepidoptera.butterflyhouse.com.au
The Department of Health: www.health.gov.au
Fishbase: www.fishbase.org
Fishes of Australia: www.fishesofaustralia.net.au
Marine Explorer: www.marineexplorer.org
Museum and Art Gallery of the Northern Territory: www.magnt.net.au
Museums Victoria: www.museumsvictoria.com.au
Nature 4 You: www.wildlifedemonstrations.com
Perth Museum: www.museum.wa.gov.au
Queensland Museum: www.qm.qld.gov.au
St John: www.stjohn.org.au
South Australian Museum: www.samuseum.sa.gov.au
Taronga Zoo: www.taronga.org.au/conservation
Tasmanian Museum and Art Gallery: www.tmag.tas.gov.au
University of Sydney, Dept. of Medical Entomology: www.medent.usyd.edu.au
Wildlife Tourism Australia: www.wildlifetourism.org.au

PUBLICATIONS

Allen, G. R. (2009). *Field Guide to Marine Fishes of Tropical Australia and South-east Asia* (4th edn). WA Museum.

Ardelean, A. & Fautin, D. (2004). Variability in nematocysts from a single individual of the sea anemone *Actinodendron arboreum* (Cnidaria: Anthozoa: Actiniaria). *Coelenterate Biology 2003: Trends in Research on Cnidaria and Ctenophora*, pp. 189–197.

Ardelean, A. & Fautin, D. (2004). A new species of the sea anemone *Megalactis* (Cnidaria: Anthozoa: Actiniaria: Actinodendridae) from Taiwan and designations of a neotype for the type species of the genus. *Proceedings of the Biological Society of Washington* 117 (4): 488–504.

ATSB Transport Safety Report (2017). *Australian Aviation Wildlife Strike Statistics.* Australian Transport Safety Bureau, ACT.

Bahrami, Y., Zhang, W. & Franco, C. (2014). Discovery of novel saponins for the viscera of the Sea Cucumber *Holothuria lessoni. Mar Drugs* 12(5): 2633–2667.

Barker, S. C., Walker, A. R. & Campelo, D. (2014). A list of the 70 species of Australian ticks; diagnostic guides to and species accounts of *Ixodes holocyclus* (paralysis tick), *Ixodes cornuatus* (southern paralysis tick) and *Rhipicephalus australis* (Australian cattle tick); and consideration of the place of Australia in the evolution of ticks with comments on four controversial ideas. *International Journal of Parasitology* 44(12): 941–953.

Bane, V., Lehane, M., Dikshit, M., O'Riordan, A. & Furey, A. (2014). Tetrodotoxin: chemistry, toxicity, source, distribution and detection. *Toxins (Basal)* 6(2): 693–755.

Bentlage, B., Cartwright, P., Yanagihara, Q. Q., Lewis, C., Richards, G. S. & Collins, A. G. (2009). Evolution of box jellyfish (Cnidaria: Cubozoa), a group of highly toxic invertebrates. *Proceedings of the Royal Society B: Biological Sciences* 277 (1680), 493–501.

Chaiwong, T., Srivora, T., Sueabsamran, P., Sukontason, K., Sanford, M. R. & Sukontason, K. L. (2014). The blow fly, *Chrysomya megacephala*, and the house fly, *Musca domestica*, as mechanical vectors of pathogenic bacteria in Northeast Thailand. *Tropical Biomedicine* 31(2): 336–46.

Chan, Thomas Y. K. (2017). Regional variations in the risk and severity of ciguatera caused by eating moray eels. *Toxins* 9 (7): 201.

Chung, J. J., Ratnapala, L. A., Cooke, I. M. & Yanagihara, A. A. Partial purification and characterization of a hemolysin (CAH1) from Hawaiian box jellyfish (*Carybdea alata*) venom. *Toxicon* 2001;39:981–990.

Cogger, H. G. (2014). *Reptiles and Amphibians of Australia* (7th edn). CSIRO, Sydney.

CrocBITE *Worldwide Crocodilian Attack Database* (accessed 18 December 2017). Charles Darwin University.

Crowther, A. (2013). Character evolution in light of phylogenetic analysis and taxonomic revision of the zooxanthellate sea anemone families Thalassianthidae and Aliciidae. *Kansas University Scholarworks*.

Dautov, S. S. & Dautova, T. N. (2016). The larvae of *Diadema setosum* (Leske, 1778) (Camarodonta: Diadematidae) from South China Sea. *Invertebrate Reproduction and Development* 60 (4): 290–296.

Deeds, J. R., Handy, S. M., White, K. D. & Reimer J. D. (2011). Palytoxin found in *Palythoa* sp. zoanthids (Anthozoa, Hexacorallia) sold in the home aquarium trade. *PLoS ONE* 6(4): e18235. doi.org/10.1371/journal.pone.0018235.

Doggett, S. & Russell, R. C. (2008). The resurgence of bed bugs, *Cimex* spp. (Hemiptera: Cimicidae) in Australia. *Proceedings of the Sixth International Conference on Urban Pests*.

Doyle, T. K., Headlam, J. L., Wilcox, C. L., MacLoughlin, E. & Yanagihara, A. A. (2017). Evaluation of *Cyanea capillata* sting management protocols using ex vivo and in vitro envenomation models. *Toxins* 9 (7): 215 DOI: 10.3390/toxins9070215.

Eipper, S. (2012). *A Guide to Australian Frogs in Captivity*. Australian Reptile Keeper Publications, Tweed Heads, NSW.

Eipper, S. (2012). *A Guide to Australian Snakes in Captivity: Elapids and Colubrids*. Australian Reptile Keeper Publications, Tweed Heads, NSW.

Fisher, M. (2005). Funnel-web spider bite: a systematic review of recorded clinical cases. *Medical Journal of Australia* 182 (8): 407–411.

Gershwin, L. & Ekins, M. (2016). A new pygmy species of box jellyfish (Cubozoa: Chirodropida) from sub-tropical Australia. *Marine Biodiversity Records* doi:10.1017/

S175526721500086X; vol. 8; e111; 2015.

Gershwin, L. & Alderslade, P. (2006). *Chiropsella bart* n. sp., a new box jellyfish (Cnidaria: Cubozoa: Chirodropida) from the Northern Territory, Australia. *The Beagle, Records of the Museums and Art Galleries of the Northern Territory*, 2006 (22).

Gray, Michael R. (2010). A revision of the Australian funnel-web spiders (Hexathelidae: Atracinae). *Records of the Australian Museum* 62 (3): 285–392.

Greer, A. E. (1997). *The Biology and Evolution of Australian Snakes*. Surrey Beatty and Sons, Chipping Norton, NSW, Australia.

Gupta, R. C. (2015). *Handbook of Toxicology of Chemical Warfare Agents* (2nd edn). Academic Press, Tokyo.

Hajdu, E. & Van Soest, R. W. M. (2002). Family Desmacellidae Ridley and Dendy, 1886, in *System Porifera: A Guide to the Classification of Sponges* [Hooper, J. N. A. & Sest, R. W. M. (eds)]. Kluwer Academic/Plenum Publishers, New York.

Hale, G. (1999). The classification and distribution of the class Scyphozoa. *Biological Diversity*, University of Oregon.

Hall, C., Levy, D. & Sattler, S. (2015). Case report: a case of palytoxin poisoning in a home aquarium enthusiast and his family. *Case Reports in Emergency Medicine*, vol. 2015, Article ID 621815.

Hartman, W. D. (1967). Revision of *Neofibularia* (Porifera Demosponginae), a genus of toxic sponges from the West Indies and Australia. *Postilla* 113: 1–43.

Hoeksema, B. W. & Crowther, A. L. (2011). Masquerade, mimicry and crypsis of the polymorphic sea anemone *Phyllodiscus semoni* and its aggregations in South Sulawesi. *Contributions to Zoology* 80 (4): 251–268.

Ihama, Y., Fukasawa, M., Kenji, N., Kawakami, Y., Nagai, T., Fuke, C. & Miyazaki, T. (2014). Anaphylactic shock caused by sting of crown-of-thorns starfish (*Acanthaster planci*). *Forensic Science International* 236, e5-e8.

Isbister, G., Gray, M., Balit, C., Raven, R., Stokes, B., Porges, K., Tankel, A., Turner, E., White, J. & Fisher, M. J. (2005) Funnel-web spider bite: a systematic review of recorded clinical cases. *Medical Journal of Australia* 182 (8): 407-411.

IUCN (2017). *The IUCN Red List of Threatened Species*. Version 2017-3. www.iucnredlist. org. Downloaded on 5 December 2017.

Jackson, S. & Groves, C. (2015). *Taxonomy of Australian Mammals*. CSIRO Publishing, Clayton South, Victoria.

Jereb, P., Roper, C., Norman, M. & Finn, J. (2016). Cephalopods of the world: an annotated and illustrated catalogue of the cephalopod species known to date. *FAO Species Catalogue for Fishery Purposes* no. 4, vol. 3.

Jereb, P. & Roper, C. F. E. (2005). Cephalopods of the world: an annotated and illustrated catalogue of the cephalopods species known to date. *FAO Species Catalogue for Fishery Purposes* no. 4, vol. 1.

Keesing, J. F., Strzlecki, J., Stowar, M., Wakeford, M., Miller, K. J., Gershwin, L. A. & Liu, D. (2016). Abundant box jellyfish, *Chironex* sp. (Cnidaria: Cubozoa: Chirodropidae), discovered at depths of over 50m on Western Australian coastal reefs. *Scientific Reports* 6, article number: 22290.

Last, P., Naylor, G., Seret, B., White, W., Stehmann, M. & Carvahlo, M. de (2016). *Rays of the World*. CSIRO Publishing, Vic.

Last, P. & Stevens, J. D. (2009). *Sharks and Rays of Australia*. CSIRO Publishing, Vic.

Lee, A. & Yanagihara, A. A. Insights into the mechanistic basis of the Irukandji Syndrome by evaluating the hematologic and immunologic responses in whole blood 2011. hdl. handle.net/10125/29642.

Manasria, T., Moussa, F., El-Hamza, S., Tine, S., Megri, R. & Chenchouni, H. (2014). Bacterial load of German cockroach (*Blattella germanica*) found in hospital environment. *Pathogens and Global Health* 108(3): 141–147.

Marsh, L. M. & Slack-Smith, S. M. (2010). *Field Guide to Sea Stingers and Other Venomous and Poisonous Marine Invertebrates of Western Australia*. Western Australian Museum, Perth, Western Australia.

Menkhorst, P. & Knight, F. (2004). *A Field Guide to the Mammals of Australia* (2nd edn). Oxford, South Melbourne, Victoria.

Mirtschin, P. J., Rasmussen, A. R. & Weinstein, S. A. (2017). *Australia's Dangerous Snakes, Identification, Biology and Envenoming*. CSIRO Publishing, Clayton South, Victoria.

Motomura, H., Last, P. & Gomon, M. (2006). A new species of the scorpionfish genus *Maxillicosta* from the coast of Australia, with a redescription of M. *whitleyi* (Scorpaeniformes: Neosebastidae). *Copeia* 2006(3): 445–459.

Motomura, H., Struthers, C. D., McGrouther, M. A. & Stewart, A. L. (2011). Validity of *Scorpaena jacksoniensis* and a redescription of S. *cardinalis*, a senior synonym of S. *cookii* (Scorpaeniformes: Scorpaenidae). *Ichthyological Research* 58: 315–332.

O'Hara, T. & Byrne, M. (2017). *Australian Echinoderms: Biology, Ecology and Evolution*. CSIRO Publishing, Vic.

Pereira, P. & Seymour, J. E. (2013). In vitro effects on human heart and skeletal cells of the venom from two cubozoans, *Chironex fleckeri* and *Carukia barnesi*. *Toxicon* 76: 310–315.

Pitt, K. A. & Lucas, C. H. (2013). Jellyfish blooms. *Springer Science and Business Media*, 4 November 2013.

Pizzey, G. & Knight, F. (Menkhorst, P., ed.) (2003). *The Field Guide to the Birds of Australia* (7th edn). Harper Collins Publishing, Sydney, New South Wales.

Puillandre N., Duda, T. F., Meyer C., Olivera, B. M. & Bouchet, P. (2015). One, four or 100 genera? A new classification of the cone snails. *Journal of Molluscan Studies* 81(1): 1–23.

Rentz, D. (2014). *A Guide to Cockroaches of Australia*. CSIRO Publishing, Vic.

Rowland, P. & Farrell, C. (2017). *A Naturalist's Guide to the Mammals of Australia*. John Beaufoy Publishing, Oxford.

Scanlon, J. D (2003). The Australian elapid genus *Cacophis*: morphology and phylogeny of rainforest crowned snakes. *Herpetological Journal* 13: 1–20.

Seiverling, E. V., Khalsa, A. & Ahrns, H. T. (2014). Pruritis and palpable purpura from leeches in the Australian Rainforest. *Elsevier IDCases* (1): 9–11.

Shine, R. (1991). *Australian Snakes – a Natural History*. Reed Books, Balgowlah, New South Wales.

Southcott, R. V. & Coulter, J. R. (1971). The effects of the southern Australian marine stinging sponges, *Neofibularia mordens* and *Lissoden-doryx* sp. *Medical Journal of Australia*

1971, pp. 895–901.

Sutherland, S. K. & Tibballs, J. (2001). *Australian Animal Toxins* (2nd edn). Oxford University Press, Melbourne.

Taronga Zoo, *Australian Shark Attack File* (accessed 14 December 2017). taronga.org.au/conservation/conservation-science-research/australian-shark-attack-file.

Tibballs, J., Yanagihara, A. A., Turner, H. & Winkel, K. (2011). Immunological and toxicological responses to jellyfish stings. *Inflammation and Allergy Drug Targets*. 2011 Oct 1;10(5): 438–446. PMCID: PMC3773479.

Underhill, D. (1990). *Australia's Dangerous Creatures*. Reader's Digest, Sydney, New South Wales.

Valdés, A. & Campillo, O. A. (2004). Systematics of pelagic aeolid nudibranchs of the family Glaucidae (Mollusca, Gastropoda). *Bulletin of Marine Science* 75(3): 381–389.

Van der Merwe, D. (2014). Freshwater cyanotoxins, in *Biomarkers in Toxicology* [Gupta, R. C., ed.]. Academic Press.

Van Dyck, S., Gynther, I. & Baker, A. [eds] (2013). *Field Companion to the Mammals of Australia*. New Holland. Chatswood, New South Wales.

Webb, C., Doggett, S. & Russell, R. (2016). *A Guide to Mosquitoes of Australia*. CSIRO Publishing, Vic.

West, J. G. (2015). *A Review of Shark Attacks in the Sydney Region*. Taronga Zoo.

West, J. G. (2011). Changing patterns of shark attacks in Australian waters. *Marine and Freshwater Research* 62, 744–754.

White, J. (2013). *A Clinician's Guide to Australian Venomous Bites and Stings*. BioCSL, Parkville, Melbourne.

Wilcox, C. L. & Yanagihara, A. A. (2016). Heated debates: hot-water immersion of ice packs as first aid for cnidarian envenomations? *Toxins* 8 (4): 97 DOI: 10.3390/toxins8040097.

Wilcox, C. L., Headlam, J. L, Doyle, T. K. & Yanagihara, A. A. (2017). Assessing the efficacy of first-aid measures in *Physalia* sp. envenomation, using solution and blood agarose-based models. *Toxins* 2017, 9 (5), p. 149.

Wilson, S. K. & Swan, G. (2017). *A Complete Guide to Reptiles of Australia* (5th edn). New Holland, Chatswood, Sydney.

Winkel, K. D., Hawdon, G. M., Ashby, K. & Ozanne-Smith, J. (2002). Eye injury after jellyfish sting in temperate Australia. *Wilderness and Environmental Medicine* 13: 203–205.

Wishart, G. (2006). Trapdoor spiders of the genus *Misgolas* (Mygalomorphae: Idiopidae) in the Sydney region, Australia, with notes on synonymies attributed to M. *rapax*. *Records of the Australian Museum* (2006) vol. 58: 1–18.

Yanagihara, A. A. & Shohet, R. V. (2012). Cubozoan venom-induced cardiovascular collapse is caused by hyperkalemia and prevented by zinc gluconate in mice. *PLoS One* 7 (12) e51368.

Yanagihara, A. A., Wilcox, C., King, R., Hurwitz, K. & Castelfranco, A. M. (2016). Experimental assays to assess the efficacy of vinegar and other topical first-aid approaches on cubozoan (*Alatina alata*) tentacle firing and venom toxicity. *Toxins* (Basel) 2016:8(1). pii: E19. doi: 10.3390/toxins8010019.

Yanagihara, A. A., Wilcox, C., Smith, J. & Surrett, G. W. (2016). Cubozoan envenomations: clinical features, pathophysiology and management, in Goffredo, S. & Dubinsky, Z. (eds), *The Cnidaria, Past, Present and Future. The World of Medusa and Her*

Sisters. Springer International Publishing, Switzerland, pp. 637–652.

Yanagihara, A. A. & Wilcox, C. (2017). Cubozoan sting-site seawater rinse, scraping, and ice can increase venom load: upending current first aid recommendations. *Toxins* 2017, 9 (3), 105.

Yoshimoto, C. M. & Yanagihara, A. A. (2002). Cnidarian (coelenterate) envenomations in Hawai'i improve following heat application. *Transactions of the Royal Society for Tropical Medicine and Hygiene* 2002:96: 300–303.

Zborowski, P. & Edwards, T. (2007). *A Guide to Australian Moths*. CSIRO Publishing, Vic.

ACKNOWLEDGEMENTS

Peter Rowland and Scott Eipper firstly thank their wives, Kate and Tie, for their wonderful and unconditional support and assistance during the writing of this book and their extended periods of research, both at home and in the field. We love you dearly.

Special thanks to Luke Allen, Shane Black, Brian Bush, Nathan Clout, Dr Harold Cogger, Kane Durrant, Euan Edwards, Colin Eipper, Adam Elliott, Dr Julian Finn, Ryan Francis, Dr Bryan Fry, Dr Lisa-Ann Gershwin, Dr John Hooper, Dr Pat Hutchings, Dr Stephen Keable, Johnny Smith Larsen, Nathan Litjens, Mark Magrouther, Ross McGibbon, Angus McNab, Mechela Mitchell, Susan Morrison, Dr Tim O'Hara, Dr Mandy Reid, Thomas Rowland, Dr Glenn Shea, Jason Sulda, Michael Swan, David Taylor, Stephen Thurstan, Jan Watson, Rachel Whitlock, Brad Whittard, Steve Wilson, Dr Ken Winkel, Nick Volpe, Dr Angel Yanagihara and Anders Zimny. Each of you was invaluable in many ways in the production of this book, whether generously reviewing and commenting on sections of the text, giving assistance in the field, supplying the scientific resources required for accuracy and currency of text content, providing contact details for associate researchers, or suggesting potential sources for many of the images that were difficult to obtain elsewhere.

Thanks to the individual photographers who either gave or otherwise made available their wonderful images, without which this style of book would not be be possible to produce, or adequately display the dramatic beauty of the species it contains. The hours of patience in the field to capture the shots is gratefully acknowledged and respected. Some have given images freely and others for greatly reduced fees, which epitomizes the generosity and common desire of the wildlife community as a whole to make the public more aware of the wondrous animals we share our planet with. While the list of image contributors is too extensive to include here, special mention goes to Ron DeCloux, Bernard DuPont, Damien Esquerre, Ryan Francis, Lisa-Ann Gershwin, Angus McNab, Sue & Rob Peatling and John Turnbull.

Additionally, thanks go to the field researchers, lab technicians and other associates that make their amazing work available in the public domain, without which authors such as ourselves would not have access to accurate, relevant and current information on the species portrayed.

Lastly, we thank the publishers, including John Beaufoy and Rosemary Wilkinson for the opportunity to write this work, series editor Krystyna Mayer for ensuring readability and consistency in the text, and Sally Bird of Calidris Literary Agency for the establishment and management of these valuable relationships.

▪ INDEX ▪